Unwin Studies in Physics and Applied Mathematics

Edited by

JOHN M. CHARAP, B.A., M.A., PH.D.
Reader in Theoretical Physics
Queen Mary College
University of London

No. 2

QUANTUM MECHANICS
AN INTRODUCTION

Unwin Studies in Physics

no. 1 DIGITAL SYSTEMS LOGIC AND CIRCUITS
B. Zacharov
Daresbury Nuclear Physics Laboratory

no. 3 VECTOR ANALYSIS
L. Marder
University of Southampton

in preparation

ELECTROMAGNETISM (2 vols.)
F. F. Heymann
Professor, University College, London

SCATTERING THEORY
S. Zienau
University College, London

QUANTUM MECHANICS
AN INTRODUCTION

J. G. Taylor

Professor of Theoretical Physics
University of Southampton

London. GEORGE ALLEN AND UNWIN LTD

FIRST PUBLISHED IN 1970

This book is copyright under the Berne Convention. All rights are reserved. Apart from any fair dealing for the purpose of private study, research, criticism or review, as permitted under the Copyright Act, 1965, no part of this publication may be reproduced, stored in a retrieval system, or transmitted, in any form or by any means, electronic, electrical, chemical, mechanical, optical, photocopying recording or otherwise, without the prior permission of the copyright owner. Enquiries should be addressed to the Publishers.

© *George Allen and Unwin Ltd 1970*

SBN 04 530014 3   *cloth*
SBN 04 530015 1   *paper*

PRINTED IN GREAT BRITAIN
in 10 on 12 pt Times type
BY WILLMER BROTHERS LIMITED
BIRKENHEAD

FOREWORD

The Series in General

The teaching of physics at universities is in a state of ferment, both in this country and elsewhere. This is in part a reflection of the activity in the subject itself, advances in research necessitating constant renewal of undergraduate syllabuses. But there is also a recognition of the need to modify the way physics is taught in order to accommodate the shifting role the subject is required to play in relation to other fields of study. Increasingly wide categories of students find they need to learn physics, or at least *some* physics. From another point of view this will mean that the students who might benefit from a particular typical course of lectures will have a variety of specialized interests and a range of different backgrounds.

An immediate consequence of this is that the traditional physics text-book may no longer be the most appropriate form for the varied class of students one may anticipate as attending a typical lecture-course. The reason for this is clear. Conventional physics text-books usually cover a much wider range of material than would normally be presented in a typical undergraduate course of lectures (say 25-35 hours). Such a book might deal with a particular sequence of ideas leading through those covered by the course. The central point is that there may be many different sequences of ideas which, as it were, converge through a particular lecture-course, and diverge on either side of it. There is of course a very important role for the sort of book which provides a synoptic view of any one of these sequences of ideas. But there is also a very real need for a series of monographs which is designed in such a way that each volume in the series covers just the material which might be treated in a single typical undergraduate lecture course. It is with this need in mind that the present series of studies in physics has been conceived.

The scope of the series then is very broad, and no topic which might reasonably be taught to undergraduate physics students is outside it. The individual contributions to the series will range in subject matter across the whole spectrum of physics, and in difficulty from introductory freshman texts to advanced monographs on particular topics such as might interest final year undergraduate students, or perhaps even postgraduate students. The unifying feature of the series will be that each volume will be course-oriented. That is to say it will be conceived as a course-text for a 25–35-hour lecture-course. And it will be written with the hope that it will be bought by the student for his own use, and not be regarded only as a reference work. As the series grows, the individual contributions will link up and a continuing effort will be made to ensure dovetailing of the joints between one contribution and its neighbours.

This Volume in Particular

Quantum mechanics is one of the pillars of modern theoretical physics, and there can be no true understanding of the phenomena of physics, or indeed of chemistry, without a prior appreciation of the quantum behaviour of atomic systems. The gadgets of modern technology from transistors to lasers exhibit essentially non-classical properties. So it is necessary in any curriculum for the teaching of physics to introduce the student to the basic principles and methods of quantum mechanics as early as is practical. Even the student of theoretical physics who will eventually demand a deeper and more mathematical approach will appreciate a prior introduction which is less formal and more directly oriented towards the physical phenomena which the theory has been developed to explain. It is the purpose of this book to provide such an introduction. For the student whose studies are directed more towards the experimental or applied aspects of physics the theoretical underpinning should still include an accurate account of quantum mechanics. This book should serve such a student well by giving him enough formal detail for him to acquire a firm foundation on which he can build further as necessary, without overburdening his understanding of the physical ideas by overemphasis of the mathematical formalities. For some students it may be felt that a single course in quantum mechanics is adequate: although this book is primarily designed as an introductory course, it has been made sufficiently self-contained to serve also as a text for a terminal, elementary course in the subject.

John M. Charap
Queen Mary College, London

CONTENTS

FOREWORD		vii
INTRODUCTION		xi

1	CLASSICAL MECHANICS	15
1.1	Introduction	15
1.2	Motion of a Single Particle	16
1.3	Angular Momentum and Generalized Co-ordinates	26
1.4	Systems of Particles	28
1.5	Variational Principles *	31
2	THE DEVELOPMENT OF WAVE MECHANICS	34
2.1	The Difficulties of Classical Mechanics	34
2.2	The Quantisation of Matter	42
2.3	Matter Waves	46
2.4	The Schrödinger Equation	50
2.5	The Postulates of Quantum Mechanics *	54
2.6	The Uncertainty Principle	65
2.7	Stationary States	70
2.8	The Motion of Wave Packets *	73
3	MOTION IN ONE DIMENSION	77
3.1	Introduction	77
3.2	The Step Potential	78
3.3	The Potential Barrier	82
3.4	The Potential Well	85
3.5	The Harmonic Oscillator	90
3.6	The General Potential	95
3.7	The Periodic Potential	97
4	MOTION IN THREE DIMENSIONS	105
4.1	Introduction	105
4.2	Angular Momentum	106
4.3	Spherically Symmetrical Potentials	118
4.4	The Hydrogen Atom	121

4.5	Time Dependence *	127
4.6	Symmetries and Conservation Laws *	130
5	**ATOMS AND MOLECULES**	139
5.1	Many Particles	139
5.2	The Hydrogen Atom	143
5.3	Perturbation Theory of Energy Levels	145
5.4	Time Dependent Perturbation Theory	147
5.5	The Zeeman Effect	151
5.6	Atoms and the Periodic Table	157
5.7	Molecules	163
6	**SCATTERING THEORY**	169
6.1	The Scattering Cross Section	169
6.2	The Scattering Amplitude	174
6.3	The Scattering Equation	176
6.4	The Born Approximation	180
6.5	Phase Shifts	183
6.6	The Square Well	186
6.7	Bound States and Resonances	189
7	**CONCLUSION**	197
7.1	Introduction	197
7.2	The Dirac Theory of the Electron	198
7.3	Whither Quantum Mechanics?	199

APPENDIX 1: Angular Momentum — 201

APPENDIX 2: Physical Constants — 203

INDEX — 205

Introduction

Mechanics attempts to explain the motions of material bodies. Classical mechanics is the attempt to explain them by means of Newton's Laws of Motion. This explanation appears to be successful provided bodies are not too small (at least larger than 10^{-10} metres, the typical atomic size) and not moving too fast (less than about one tenth the velocity of light, 3×10^8 metres per second). Most bodies met with in everyday experience satisfy these two conditions, so that classical mechanics may be regarded as the mechanics of everyday bodies.

However, classical mechanics breaks down when we discuss the properties of bodies of atomic size or less. Quantum mechanics has been developed for such bodies, the most important examples of which are atoms and molecules and their constituents. It is the purpose of this book to describe the fundamentals of quantum mechanics and the remarkable explanation it gives of atomic and molecular structure and behaviour.

The special assumptions of classical mechanics which differentiate it from quantum mechanics have now become clear, as well as the common properties of these two approaches; at the same time the demarcation line between them has become apparent. This does not mean that classical and quantum mechanics are two *different* approaches with a sharply defined dividing line between them. Although this is true in a limited sense, it would be closer to the nature of physical explanation to say that quantum mechanics is to be regarded as universally applicable to mechanical systems, but gives results indistinguishable from those of classical mechanics for the motion of macroscopic bodies. So the demarcation line between classical and quantum mechanics occurs as soon as quantum effects may be neglected, in other words for systems much larger than atoms.

This book is based on lectures given to undergraduates who are meeting quantum mechanics for the first time. It is meant to cover a one-year course for students who will not necessarily specialize in theoretical physics. The only prerequisite for understanding the book is a reasonable familiarity with calculus, including partial differentiation, and of elementary vector analysis. It is usually expected in a course of quantum mechanics that a considerable amount of classical mechanics has already been absorbed. Such mechanics may have been learnt mainly at school, and as such will not necessarily be 'on tap' when needed. I have included a brief account of the classical mechanics of a single particle and of a system of particles. Since the quantisation of only very simple classical systems are to be discussed, such an account should be satisfactory for a student whose knowledge of classical mechanics is minimal. At the same time it enables the important physical concepts such as potential, linear momentum, angular momentum, etc., to be considered in a classical frame, and the properties of these concepts to be discussed in a fashion which allows for an illuminating comparison with their quantum analogues.

The approach used in this book is to consider the quantum mechanics of a system as arising from the classical mechanics of that system by a set of well defined rules. These rules for quantising a classical system can only be made explicit after the classical system has been discussed. Hence the account of classical mechanics enables us at the same time to state our notation for such systems, so that later we can state precisely the rules for quantising them. Such an approach will evidently highlight the differences between the classical and quantal behaviour of the same system.

Since this is an introductory book, certain important topics are touched only cursorily in it, or are not touched at all. In particular, the effects arising from the motion of very fast bodies are not considered, so that the Dirac theory of the electron is excluded (except for some brief remarks in Chapter 7). However, the non-relativistic effects of electron spin are discussed, in particular the effect of the exclusion principle on atomic structure and its explanation of the periodic table of elements.

Chapter 1 contains an account of the simpler aspects of the classical mechanics of single particles and systems of particles, defining the particular dynamical variables of momentum, angular momentum, Lagrangian and Hamiltonian, and discussing under what conditions these quantities do not vary in time. Chapter 2 begins with a description of the various difficulties which faced classical mechanics in trying to explain various properties of matter. The partial explanations provided by Planck's quantum hypothesis are then described, leading to the Bohr-Rutherford model of the atom. The difficulties still unexplained by this model are mentioned; the wavelike properties of matter are described and used to introduce the Schrödinger wave function and the Schrödinger wave equation which it satisfies. The postulates of quantum mechanics are discussed in more detail in §2.5, and shown to lead to the uncertainty principle in the next section; the stability of matter is then considered, and finally the classical nature of the overall motion of wave functions considered in §2.8.

In Chapter 3 we consider a number of soluble problems of motion described by the Schrödinger equation in one dimension, culminating with a discussion of the one dimensional periodic potential and the nature of electronic motion in solids. Chapter 4 discusses motion in three dimensions. We consider first the angular momentum operator in quantum mechanics, and discuss how it is a constant in time and so may be used to label states when the potential is spherically symmetric. We then obtain the energy levels of the hydrogen atom, so finally explaining the frequencies of spectral lines. Lastly we consider the manner in which symmetries (such as rotation symmetry) and conserved quantities are related, together with a general discussion of time dependence.

In Chapter 5 we consider many-particle systems, and approximation methods for obtaining their energy levels and transition rates between their levels. These methods are applied to give a detailed description of the normal Zeeman effect, and of the building up of the periodic table and the binding of atoms into molecules.

Chapter 6 discusses scattering theory, both in terms of the basic scattering and its physical meaning and in terms of the phase shift analysis. A final discussion in the chapter considers the way bound state and scattering properties are related, and how decaying 'particles' or resonances occur which possess an intimate mixture of scattering and bound state properties.

The last chapter contains a brief introduction to the Dirac equation and concludes with a description of the two distinct ways in which quantum mechanics is presently being used to analyse the properties of matter.

In order to cater for the lower level of mathematical interest of some readers, we have put an asterisk next to more mathematical sections which are not essential to the later development; a student studying quantum mechanics on his own might also be advised to leave out these sections on a first reading. However, the summary of §2.5, given at the end of that paragraph, is necessary, since the ideas summarized are used constantly throughout the later development in the book.

We use SI units throughout the book. The notation is standard; in particular, vectors are denoted by bold face type, with the position vector being $\mathbf{r}$ with components (x, y, z) or (r_1, r_2, r_3), whilst the momentum vector is $\mathbf{p}$ with components (p_x, p_y, p_z) or (p_1, p_2, p_3). The length of a vector will be $|\mathbf{r}| = \sqrt{(x^2 + y^2 + z^2)}$, and is sometimes denoted by an italic r; $|\mathbf{r}| = r$. Finally indices always take the values 1 to 3, unless otherwise stated (as in §7.2).

I would like to acknowledge the great help from the editor, Dr. Charap, and also to thank Mr. Lynch-Blosse for his encouragement, and my typist, Miss Barbara Correy, for her excellent work. Finally I would like to thank my wife, Pat, for fitting this book into the wider framework of life.

CHAPTER 1

Classical Mechanics

1.1 Introduction

This first chapter is an introduction and brief survey of the salient features of classical mechanics. It is meant to enable students who have only a little familiarity with classical mechanics to understand the basic mechanical concepts of momentum, angular momentum, potential function, Hamiltonian, etc., in the classical case. These concepts are extremely important throughout the whole of mechanics, both of a classical and a quantal kind. They are also important in the development of quantum mechanics which we give in this book. As was stated in the introduction, we quantise a mechanical system by stating the quantisation rules by which we transform the classical system to a quantal one. These rules are expressed in terms of the basic mechanical quantities for the system, and a preliminary discussion of these quantities will be necessary to specify our notation and define them. This chapter will be limited to only the simplest possible systems, and mainly to that of a single particle moving in a conservative field of force. The quantum mechanics of this system and its simple extensions will be the only one we will consider in any detail. We thus give a reasonably self contained discussion of both the classical and quantum mechanics of such simple systems. Of course we cannot hope to cover the classical mechanics of complicated bodies, such as rolling discs, spinning tops, etc., etc. (The reader interested in such topics is advised to turn to the texts referred to at the end of the chapter.)

1.2 Motion of a Single Particle

We will first develop the classical mechanics of a single point particle. Such an idealized point particle is a good approximation to a physical particle provided we do not consider mechanical effects which depend on the internal structure of such a particle.

The position of the particle may be described with respect to some origin of co-ordinates by a vector **r**. We denote the components of this vector in some rectangular co-ordinate system as $\mathbf{r} = (x, y, z)$. The vector **r** or its components will be functions of the time t; the actual dependence on t will determine the motion of the particle.

The basic equation of motion which describes the time development of **r** is Newton's Second Law of Motion. We assume that the particle is moving under some external force vector **F** applied to it. We also assume that the particle has a constant mass m, so the momentum **p** of the particle is

$$\mathbf{p} = m\dot{\mathbf{r}} \tag{1.1}$$

where $\dot{\mathbf{r}}$ is the velocity vector of the particle, $\dot{\mathbf{r}} = d\mathbf{r}/dt = (dx/dt, dy/dt, dz/dt)$. Newton's Second Law of Motion then states that the external force on the particle is equal to the time rate of change of its momentum, or vectorially:

$$\mathbf{F} = \dot{\mathbf{p}} = m\ddot{\mathbf{r}} \tag{1.2}$$

We expect that the force **F** on the particle will depend on where the particle is, so **F** is a known function of **r**, $\mathbf{F} = \mathbf{F}(\mathbf{r})$. The function $\mathbf{F}(\mathbf{r})$ in general depends on **r** in a complicated fashion. Then equation (1.2) becomes

$$\mathbf{F}(\mathbf{r}) = m\ddot{\mathbf{r}} \tag{1.3}$$

The basic problem of classical mechanics of a single particle becomes that of solving the second order differential equation (1.3). The solution of such an equation is given if the position **r** and the velocity $\dot{\mathbf{r}}$ of the particle are given at some time t_0, say. For then (1.3) allows the acceleration $\ddot{\mathbf{r}}$ at time t_0 to be obtained, and by differentiation of equation (1.3) with time it will also enable all the higher time derivatives of **r** to be obtained at time t_0. We can then obtain the value of **r** at a later time t by means of the Taylor expansion

$$\mathbf{r}(t) = \sum_n \frac{1}{n!} (t-t_0)^n \frac{d^n}{dt^n} \mathbf{r}(t_0)$$

Such an expansion may only converge for suitably small t, but we can perform another expansion around a later time t'_0, say, and then another around t''_0, say, and so on till the particle motion $\mathbf{r}(t)$ is obtained for all times t. This motion is called the particle path.

It is usually difficult to find the particle path explicitly from the equation of motion (1.3). Because of this it is useful to obtain some less precise details of the motion than its complete form. Such details may still enable important general features of the motion to be derived. A particularly important quantity which will help us to get general features is an integral of the motion. This quantity is constant along the particle path, and so is constant in the course of time. We expect that the more integrals of the motion are obtained, the more complete the description of the particle path.

One very important integral is the energy of the particle. This may be introduced when the force $\mathbf{F}$ is conservative. A conservative force is one for which the work done by the force in moving the particle round any closed path is zero, so $\int_C \mathbf{F} \cdot d\mathbf{r} = 0$, where C is any closed contour. An equivalent expression for a conservative field of force is that the work done by the force in moving the particle from any point A to any other point B is independent of the path taken by the particle from A to B, so $\int_A^B \mathbf{F} \cdot d\mathbf{r}$ is uniquely defined for any curve joining A to B. If B has the position vector $\mathbf{r}$ and A is a fixed point, then we define the function of $\mathbf{r}$ with value $-\int_A^B \mathbf{F} \cdot d\mathbf{r}$ to be the potential function $V(\mathbf{r})$:

$$\int_\mathbf{r}^A \mathbf{F} \cdot d\mathbf{r} = V(\mathbf{r}) \tag{1.4}$$

Since $\mathbf{F}$ is conservative, there is no ambiguity in the definition of V arising from various choices of path from $\mathbf{r}$ to A; all paths give the same function $V(\mathbf{r})$.

We may reverse (1.4) by differentiation:

$$\mathbf{F} = -\nabla V \tag{1.5}$$

where ∇ is the vectorial operator $\nabla = (\partial/\partial x, \partial/\partial y, \partial/\partial z)$. Evidently $V(\mathbf{r})$ depends on the fixed point A, but this dependence is trivial, since if we choose another fixed point A' in place of A, then the new potential function V', say, is related to the old one, V, by constant addition:

$$V'(\mathbf{r}) = \int_\mathbf{r}^{A'} \mathbf{F} \cdot d\mathbf{r} = \int_\mathbf{r}^A \mathbf{F} \cdot d\mathbf{r} + \int_A^{A'} \mathbf{F} \cdot d\mathbf{r} = V(\mathbf{r}) + \int_A^{A'} \mathbf{F} \cdot d\mathbf{r} \tag{1.6}$$

This constant addition $\int_A^{A'} \mathbf{F} \cdot d\mathbf{r}$ cannot affect the motion of the particle since the force $\mathbf{F}$ is determined from the gradient of V, by equation (1.5), and the gradient of a constant is zero. We see then that the potential function is not determined to within an additive constant. It is usual to fix that constant by taking A very far away; this can be done only if $\mathbf{F}$ is a decreasing function of $\mathbf{r}$ for large $\mathbf{r}$.

Let us now use the potential function to obtain an integral of the motion of a single particle moving in a conservative field of force. We evaluate the left-hand side of (1.4) by using the equation of motion (1.3):

$$\int_A^B \mathbf{F} \cdot d\mathbf{r} = \int_A^B m\ddot{\mathbf{r}} \cdot d\mathbf{r} = \int_A^B m\ddot{\mathbf{r}} \cdot \dot{\mathbf{r}} \, dt$$

$$= \int_A^B \frac{d}{dt}(\tfrac{1}{2}m\dot{\mathbf{r}}^2) \, dt = \tfrac{1}{2}m\dot{\mathbf{r}}_B^2 - \tfrac{1}{2}m\dot{\mathbf{r}}_A^2 \qquad (1.7)$$

But the left-hand side of (1.7), the work done in moving the particle from A to B is, by (1.5),

$$-\int_A^B \nabla V \cdot d\mathbf{r} = -\int_A^B dV = V(A) - V(B) \qquad (1.8)$$

Thus from (1.7) and (1.8) we obtain

$$[\tfrac{1}{2}m\dot{\mathbf{r}}_A^2 + V(A)] = [\tfrac{1}{2}m\dot{\mathbf{r}}_B^2 + V(B)] \qquad (1.9)$$

Equation (1.9) states that at any point on the path of the particle the quantity $(\tfrac{1}{2}m\dot{\mathbf{r}}^2 + V)$, evaluated at that point, is independent of the point and so is constant in time. This constant quantity is called the energy of the particle, and is denoted by E, a real number:

$$\tfrac{1}{2}m\dot{\mathbf{r}}^2 + V = E \qquad (1.10)$$

The arbitrary constant which may be added to the potential V does not affect (1.9), since any such constant will cancel from both sides. Such is not the case in (1.10), and adding a constant to V will add the same constant to E. Thus the energy E is not determined to within an additive constant. Again such an arbitrariness will not affect the particle motion.

The energy is the sum of two terms, one of which is the potential function V. This latter contribution is usually called the potential energy, whilst the quantity $\frac{1}{2}m\dot{r}^2 = \frac{1}{2}mv^2$ is called the kinetic energy, and denoted by T:

$$E = T+V \quad (1.11)$$

$$T = \tfrac{1}{2}mv^2 \quad (1.12)$$

We may rewrite E and T in terms of $\mathbf{r}$ and the momentum $\mathbf{p}$ of the particle:

$$\mathbf{p} = m\mathbf{v} = m\dot{\mathbf{r}}$$

$$E = (\mathbf{p}^2/2m) + V(\mathbf{r}) = H(\mathbf{r}, \mathbf{p}) \quad (1.13)$$

where the expression for E in terms of $\mathbf{r}$ and $\mathbf{p}$ is called the Hamiltonian for the system and denoted by H or $H(\mathbf{r}, \mathbf{p})$.

We obtain a very important limitation on possible motions if we note that $\dot{\mathbf{r}}^2$ or $\mathbf{p}^2$ must always be positive for a real motion. Thus from (1.10) or (1.13) we have

$$E - V \geqslant 0 \quad (1.14)$$

Equation (1.14) may be solved if the potential function $V(\mathbf{r})$ is known explicitly and the value of energy E given, so defining the explicit region in which classical motion can occur. Two examples follow.

Example 1: Inverse Square Law of Force. In this case the direction of $\mathbf{F}$ is parallel to that of $\mathbf{r}$, whilst the magnitude of $\mathbf{F}$ is r^{-2}, where $r = |\mathbf{r}|$. Hence

$$\mathbf{F} = -\lambda \mathbf{r}/r^3 \quad (1.15)$$

A particular example is the gravitational force acting on the particle due to a gravitating body of mass M at the origin, when $\lambda = GMm$, where G is the gravitational constant, the force being attractive. Another example is that of electrostatic interaction of two charged particles with charges e and e', when $\lambda = -ee'/4\pi\epsilon_0$ where ϵ_0 is the permittivity of free space. If we choose the potential to vanish at infinity then

$$V(\mathbf{r}) = -\lambda/r$$

(since then $-\nabla V$ gives (1.15)). Thus

$$H = (\mathbf{p}^2/2m) - \lambda/r$$

In Fig. 1.1 we draw the curve of $V(r)$ against r for $\lambda < 0$ and the region of possible motion when E is either positive or negative; Fig. 1.2 is for $\lambda > 0$.

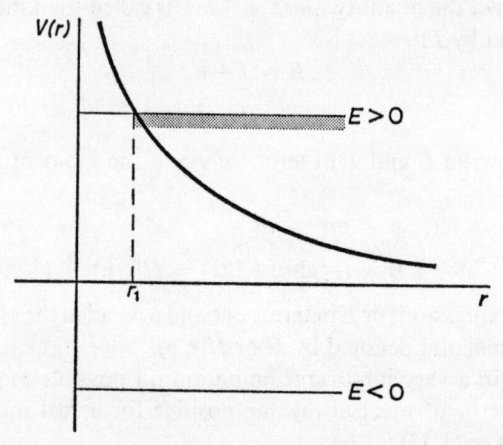

Fig. 1.1 The repulsive Coulomb potential $-\lambda/r$, $(\lambda < 0)$, plotted against the radial distance r. The shading indicates the region of possible classical motion $r > r_1$ for positive $E > 0$; there is no region of classical motion when $E < 0$. The value of r_1 is $-\lambda/E$.

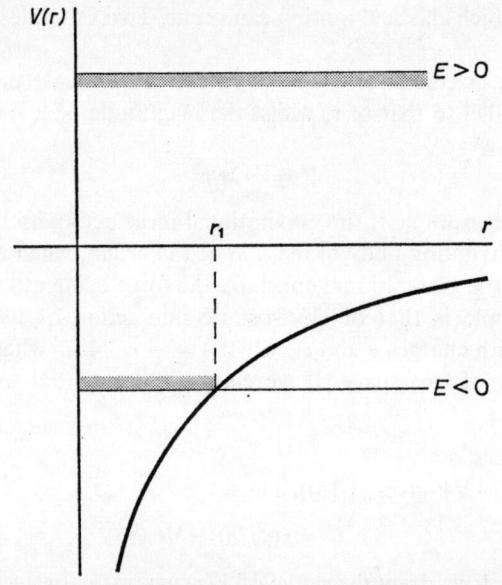

Fig. 1.2 Similar to Fig. 1.1 except the potential is attractive $(\lambda > 0)$. The possible classical motion for $E > 0$ is for all r; for $E < 0$ the possible motion is restricted to the finite region $r < r_1$.

In the first case, $\lambda < 0$, we see that classical motion is possible if $r > r_1$, where $r_1 = -\lambda/E$; if $E < 0$, then r_1 is negative so there can be no classical motion. In the second case, $\lambda > 0$, there is unlimited classical motion if $E > 0$, whilst for $E < 0$ there is only motion for $0 < r < r_1$.

Example 2: Harmonic Oscillator. In this case there is a restoring force on the particle directed to the origin of force which is proportional to the distance from this origin. Taking this origin of force to be the origin of co-ordinates, we have

$$\mathbf{F} = -\lambda \mathbf{r}$$

so that (with the point A in the definition (1.4) of potential taken as the origin)

$$V = \tfrac{1}{2}\lambda \mathbf{r}^2 \tag{1.16}$$

and

$$E = \mathbf{p}^2/2m + \tfrac{1}{2}\lambda \mathbf{r}^2 = H(\mathbf{r}, \mathbf{p}) \tag{1.17}$$

The curve of $V(r)$ against r and the region of possible motion is shown in Fig. 1.3 when $\lambda > 0$ and in Fig. 1.4 when $\lambda < 0$. In Fig. 1.3 the classical motion is the bounded region $r < r_2$, when the energy $E \geqslant 0$; when $E < 0$ there is no possible motion.

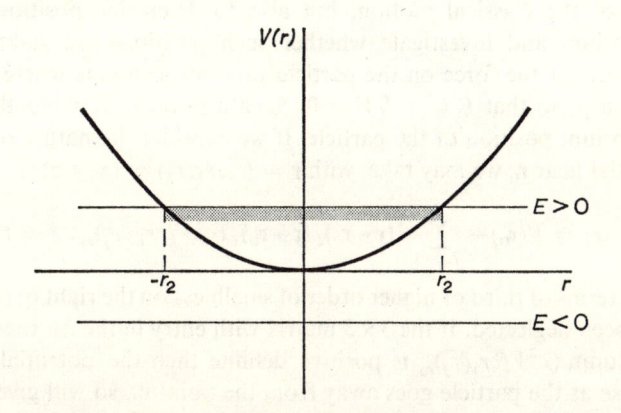

Fig. 1.3 The attractive harmonic oscillator potential $V = \tfrac{1}{2}\lambda r^2$, with $\lambda > 0$, plotted against the radial distance r (regarded as a one-dimensional variable continued to negative r as shown); the region of possible classical motion (indicated by the shading) for energy $E > 0$ is restricted to $|r| < r_2$ where $E = \tfrac{1}{2}\lambda r_2^2$. When $E < 0$ there is no classical motion.

In Fig. 1.4 the classical motion is for $r > r_2$ when $E < 0$, where $E = \frac{1}{2}\lambda r_2^2$; there is no limitation on the motion when $E > 0$. We have drawn Figs. 1.3 and 1.4 as if they involved motion for negative values of r. These figures may be interpreted in this way for one dimensional motion; the region $r < 0$ should be ignored if we are considering r as the radial distance $|\mathbf{r}|$ in three dimensional motion.

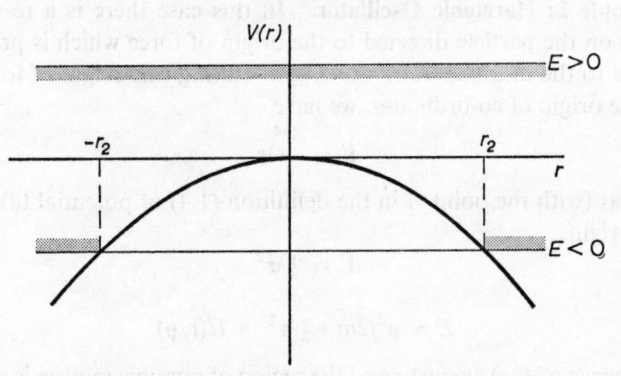

Fig. 1.4 Similar to Fig. 1.3 except the potential is repulsive, $\lambda < 0$. The possible classical motion for $E > 0$ is for all r; for $E < 0$ the classical motion is restricted to the region $|r| > r_2$.

Not only may we use the potential function V to determine the limits of the classical motion, but also to determine positions of equilibrium and investigate whether such positions are stable or unstable, for the force on the particle vanishes at points where V is stationary, so that $\mathbf{F} = -\nabla V = \mathbf{0}$. Such a point $\mathbf{r}_0$ is evidently an equilibrium position of the particle. If we consider the nature of the potential near $\mathbf{r}_0$ we may take, with $\mathbf{r} = (r_1, r_2, r_3) = (x, y, z)$

$$V(\mathbf{r}) = V(\mathbf{r}_0) + \sum_{i,j=1}^{3} \tfrac{1}{2}(\mathbf{r}-\mathbf{r}_0)_i\,(\mathbf{r}-\mathbf{r}_0)_j\,(\partial^2 V/\partial r_i\,\partial r_j)_{\mathbf{r}_0} \qquad (1.18)$$

where terms of third or higher order of smallness on the right of (1.18) have been neglected. If the 3×3 matrix with entry in the ith row and jth column $(\partial^2 V/\partial r_i \partial r_j)_{\mathbf{r}_0}$ is positive definite then the potential will increase as the particle goes away from the point $\mathbf{r}_0$, so will give rise to a restoring force. This restoring force implies stability of the equilibrium position $\mathbf{r}_0$; similarly a negative definite 3×3 matrix will give an unstable equilibrium. There will also be intermediate situations in which there will be stability for motions in certain directions away from r_0, and instability for motions in other directions. Such a possibility can arise in one dimensional motion, when stability corresponds to d^2V/dr^2 being positive at r_0, instability to d^2V/dr^2 being negative, and the mixed situation to d^2V/dr^2 being zero, when the curve of $V(r)$ against r will have a point of inflexion at r_0. The origin A is a point of stable equilibrium for the harmonic oscillator with $\lambda > 0$, as we see immediately in Fig. 1.3; A is an unstable point when $\lambda < 0$, as we see from Fig. 1.4.

So far we have discussed the motion of a particle as arising directly from Newton's Second Law of Motion, equation (1.2). There are various alternative forms of writing the equation which are especially useful for describing the motion of systems of particles. It will be helpful to consider these forms in the particularly simple case of a single particle moving in a conservative field of force.

To begin with we will consider the form of the equation of motion called *Hamilton's equations*. These are obtained from the Hamiltonian $H(\mathbf{r}, \mathbf{p})$ given in equation (1.13). We consider the partial derivatives of H with respect to the components of $\mathbf{r}$ and $\mathbf{p}$:

$$\partial H/\partial p_j = p_j/m = v_j \tag{1.19}$$

$$\partial H/\partial r_j = \partial V/\partial r_j = -F_j \tag{1.20}$$

We now use the definition of velocity $\mathbf{v} = \dot{\mathbf{r}}$ in (1.19) and Newton's Second Law (1.2) in the form $\mathbf{F} = \dot{\mathbf{p}}$ in (1.20), to obtain

$$\dot{r}_j = \frac{\partial H}{\partial p_j}; \quad \dot{p}_j = -\frac{\partial H}{\partial r_j} \tag{1.21}$$

which are Hamilton's equations of motion. These equations are equivalent to Newton's Second Law, as we see by reversing the steps leading from (1.2) to (1.21), using the explicit form of H in (1.13). They form a set of first order differential equations for the time development of the position vector *and* momentum vector, in contrast to Newton's Second Law (1.2), which is a second order differential equation for the position vector alone. This is to be expected; the reduction of a set of second order differential equations to first order equations always produces twice as many equations.

Another important form of the equations of motion derives from *the Lagrangian*. This is a function $L(\mathbf{r}, \dot{\mathbf{r}})$ which depends on $\mathbf{r}$ and $\dot{\mathbf{r}}$, and from which Newton's Law (1.2) can be obtained in a way similar to that from H, but now in terms of $\mathbf{r}$ and $\dot{\mathbf{r}}$ and not $\mathbf{r}$ and $\mathbf{p}$. The Lagrangian is defined by

$$L(\mathbf{r},\dot{\mathbf{r}}) = \dot{\mathbf{r}} \cdot \mathbf{p} - H(\mathbf{r},\mathbf{p}) \tag{1.22}$$

where $\mathbf{p}$ is to be replaced by $m\dot{\mathbf{r}}$ wherever it appears on the right hand side of (1.22). Performing this replacement we obtain

$$L = \tfrac{1}{2}m\dot{\mathbf{r}}^2 - V$$

If we now differentiate L partially with respect to $\mathbf{r}$ and $\dot{\mathbf{r}}$ we obtain

$$\left.\begin{array}{l} \partial L/\partial r_j = -\partial V/\partial r_j = F_j \\ \partial L/\partial \dot{r}_j = m\dot{r}_j = p_j \end{array}\right\} \tag{1.23}$$

We now use Newton's Second Law to combine the two equations (1.23):

$$\frac{d}{dt}\left(\frac{\partial L}{\partial \dot{r}_j}\right) = \frac{\partial L}{\partial r_j} \tag{1.24}$$

These are the *Lagrange equations of motion*. Whilst for a single particle these equations are precisely the same as Newton's Second Law, we will find that for a system of particles forming a rigid body they are more extensive. The Lagrangian function is important in solving the motion of complicated systems and also in setting up the equations of motion for new systems by analogy.

Example 3: We will consider, as an example of the use of a Hamiltonian, the derivation of the equations of motion for a charged particle moving with velocity $\mathbf{v}$ in an electric field $\mathbf{E}$ and a magnetic field $\mathbf{B}$. These fields are derived from a scalar potential ϕ and vector potential $\mathbf{A}$ as*

$$\mathbf{B} = \nabla \wedge \mathbf{A} \tag{1.25}$$

$$\mathbf{E} = -\nabla\phi - \partial \mathbf{A}/\partial t \tag{1.26}$$

Let e be the charge on the particle. Then we will try to prove that the Hamiltonian H for the charged particle is obtained by replacing $\mathbf{p}$ by $\mathbf{p} - e\mathbf{A}$ in the free particle Hamiltonian (1.17) and adding $e\phi$ to it. In the case of no other external force we have

$$H = (\tfrac{1}{2}m)(\mathbf{p} - e\mathbf{A})^2 + e\phi \tag{1.27}$$

*$\mathbf{A} \wedge \mathbf{B}$ denotes the vector product of $\mathbf{A}$ and $\mathbf{B}$, so that if $\mathbf{A} = (A_1, A_2, A_3)$, $\mathbf{B} = (B_1, B_2, B_3)$, then $\mathbf{A} \wedge \mathbf{B} = (A_2 B_3 - A_3 B_2, A_3 B_1 - A_1 B_3, A_1 B_2 - A_2 B_1)$.

We wish to show that the equation of motion derived from (1.27) is Newton's Second Law of Motion, with the force acting on the particle equal to the Lorentz force, $e\mathbf{E} + e\mathbf{v} \wedge \mathbf{B}$, so that

$$m\ddot{\mathbf{r}} = e\mathbf{E} + e\mathbf{v} \wedge \mathbf{B} \quad (1.28)$$

To derive (1.28) from the Hamiltonian (1.27) we use the Hamiltonian equations (1.21). Then differentiating (1.27) following (1.21) we obtain

$$\dot{r}_j = \partial H/\partial p_j = (p_j - eA_j)/m \quad (1.29)$$

$$-\dot{p}_j = \partial H/\partial r_j = e(\partial \phi/\partial r_j) - (e/m)\sum_{l=1}^{3}(p_l - eA_l)\,\partial A_l/\partial r_j \quad (1.30)$$

Differentiating (1.29) and substituting p_j from (1.30) we finally obtain

$$m\ddot{r}_j = \dot{p}_j - e\frac{dA_j}{dt}$$

$$= \dot{p}_j - e\left(\frac{\partial A_j}{\partial t} + \sum_{k=1}^{3}\frac{\partial A_j}{\partial r_k}\dot{r}_k\right)$$

$$= -e\frac{\partial \phi}{\partial r_j} - e\frac{\partial A_j}{\partial t} - e\sum_{k=1}^{3}\dot{r}_k\frac{\partial A_j}{\partial r_k} + \frac{e}{m}\sum_{k=1}^{3}\left(\frac{\partial A_k}{\partial r_j}\right)(p_k - eA_k) \quad (1.31)$$

Substituting for the first two terms in the last line of (1.31) by (1.26) and the last term of that line by (1.29), and then using (1.25), we obtain

$$m\ddot{r}_j = eE_j - e\sum_{k=1}^{3}\dot{r}_k\{(\partial A_j/\partial r_k) - (\partial A_k/\partial r_j)\}$$

$$= eE_j + e(\mathbf{v} \wedge \mathbf{B})_j$$

which is the jth component of the required equation of motion (1.28).

A particular case of this is for an electron of charge $-e$ moving round a nucleus of charge Ze, so that $\mathbf{A} = 0$; the last term in (1.27) is now $V = -e\phi = -Ze^2/4\pi\varepsilon_0 r$, and

$$H = (\mathbf{p}^2/2m) - Ze^2/(4\pi\varepsilon_0 r)$$

This potential has already been discussed in example 1, and the curve of V drawn in Fig. 1.2 (since $\lambda = Ze^2/4\pi\varepsilon_0$ is positive). We see from this figure that if the energy of the electron is positive, $E > 0$, then all of space is accessible to it; if $E < 0$, only a limited region $r < r_0 = Ze^2/4\pi\varepsilon_0|E|$ is accessible to the electron. Such a negative energy electron will be a bound electron, whilst the electron with energy $E > 0$ will be a free electron. We will see in the next chapter that a bound electron certainly cannot be described by such a classical picture, but the quantum mechanical description of a bound electron will also have $E < 0$ and be localized in a similar fashion.

1.3 Angular Momentum and Generalised Co-ordinates

The moment of the momentum of a single particle about a point O, usually called the angular momentum, is another important quantity both in classical and quantum mechanics. For a single particle moving classically this angular momentum (about the origin) is

$$\mathbf{L} = \mathbf{r} \wedge \mathbf{p} = \mathbf{r} \wedge m\dot{\mathbf{r}} \tag{1.32}$$

with components

$$L_x = (yp_z - zp_y),\ L_y = (zp_x - xp_z),\ L_z = (xp_y - yp_x).$$

When the particle moves according to Newton's Second Law of Motion (1.2), the time rate of change of $\mathbf{L}$ is

$$\dot{\mathbf{L}} = m\frac{d}{dt}(\mathbf{r} \wedge \dot{\mathbf{r}}) = m(\dot{\mathbf{r}} \wedge \dot{\mathbf{r}} + \mathbf{r} \wedge \ddot{\mathbf{r}}) = m\mathbf{r} \wedge \ddot{\mathbf{r}} = \mathbf{r} \wedge \mathbf{F} \tag{1.33}$$

Thus the rate of change of angular momentum about O is equal to the moment of the external force on the particle about O.

There are two important cases in which $\mathbf{L}$ is constant in time. These are

(a) if there is no external force, $\mathbf{F} = \mathbf{0}$, or
(b) if the external force is directed towards (or away from) O, so that $\mathbf{F}$ is parallel to the radius vector $\mathbf{r}$, and $\mathbf{r} \wedge \mathbf{F} = \mathbf{0}$.

In these two cases the angular momentum $\mathbf{L}$ is constant in time. Thus it is a conserved quantity, and is of great importance in specifying the motion, at least partially. Hence it would be helpful to set up Newton's Second Law of Motion so that angular momentum appears in the same manner as linear momentum; this can be achieved by introducing *generalized co-ordinates*. As their name indicates, such co-ordinates are a generalization of rectangular co-ordinates, which are needed to specify the position of a particle, either to other types of

CLASSICAL MECHANICS

(possibly non-orthogonal) co-ordinates, such as polar co-ordinates, or to co-ordinates needed to describe the position of a rigid body.

Suppose such generalized co-ordinates are denoted by q_j (j taking the values 1, 2, $\cdots$, N, where there are N such co-ordinates). We suppose that the kinetic energy T may be calculated as a function of the q_j's and $\dot{q}_j$'s, and that the motion is in a conservative field of force with potential function V depending on the q_j's. Following (1.22) we define the Lagrange function $L = T - V$, and with (1.23) introduce generalized momentum p_j with $p_j = \partial L/\partial \dot{q}_j$. We may now consider T as a function of the q_j's and p_j's, and with $H = T + V$ the Hamiltonian equations of motion are expected to be (in analogy with (1.21))

$$\dot{q}_r = \partial H/\partial p_r, \qquad \dot{p}_r = -\partial H/\partial q_r \qquad (1.34)$$

and similarly the Lagrange equations of motion (in analogy with (1.24)) are expected to be

$$\frac{d}{dt}\left(\frac{\partial L}{\partial \dot{q}_r}\right) - \frac{\partial L}{\partial q_r} = 0 \qquad (1.35)$$

We will derive (1.34) and (1.35) in the next section; let us for the moment assume that they are true. We wish to give an example which shows that when we take q_r to be an angle then the related momentum p_r is an angular momentum.

Example 4: We consider a single particle of mass m moving in a plane under a conservative force with potential $V(r)$, where r is the distance to the origin O of co-ordinates. Let P be the position of the particle at time t and θ the angle that the line OP makes with some fixed line. Then the velocity of the particle is $\dot{r}$ along OP and $r\dot{\theta}$ perpendicular to OP, so that the particle's kinetic energy is $T = \frac{1}{2}m(\dot{r}^2 + r^2\dot{\theta}^2)$. Defining $L = T - V$, and regarding r and θ as the generalised co-ordinates describing the motion of the particle, the corresponding momenta p_r, p_θ will be

$$p_r = \partial L/\partial \dot{r} = m\dot{r}, \qquad p_\theta = \partial L/\partial \dot{\theta} = mr^2\dot{\theta} \qquad (1.36)$$

Evidently p_θ is the angular momentum of the particle about O. We will determine the Hamiltonian equations from $H = T + V$, writing H as a function of the co-ordinates r, θ and the momenta p_r, p_θ:

$$H = p_r^2/2m + p_\theta^2/2mr^2 + V(r) \qquad (1.37)$$

The equations (1.34) become:

$$\dot{r} = \partial H/\partial p_r = p_r/m \qquad \text{(agreeing with (1.36))}$$

$$\dot{\theta} = \partial H/\partial p_\theta = p_\theta/mr^2 \qquad \text{(agreeing with (1.36))}$$

$$\dot{p}_\theta = \partial H/\partial \theta = -\partial V/\partial \theta \qquad (1.38)$$

$$\dot{p}_r = \partial H/\partial r = p_\theta^2/mr^3 - \partial V/\partial r \qquad (1.39)$$

We may interpret the last term in (1.38) as the couple G of the external force about O, since the work done in a small rotation $d\theta$ by a couple is $Gd\theta$, but is also equal to $-dV = -(\partial V/\partial \theta)\, d\theta$. Thus $G = -\partial V/\partial \theta$, and (1.38) is the angular analogue of Newton's Second Law of Motion (1.2). If we now insert p_r and p_θ from (1.36) into the left-hand sides of (1.38) and (1.39) we have

$$\frac{d}{dt}(mr^2\dot{\theta}) = -\frac{\partial V}{\partial \theta}, \qquad m\dot{r} = mr\dot{\theta}^2 - \frac{\partial V}{\partial r} \qquad (1.40)$$

If, further, the external force **F** is along **r** (the vector OP), then V is independent of θ, and (1.40) becomes

$$mr^2\dot{\theta} = \text{constant} \qquad (1.41)$$

$$m(\ddot{r} - r\dot{\theta}^2) = -\partial V/\partial r \qquad (1.42)$$

(1.41) is the statement of conservation of angular momentum about O, (1.42) is the radial form of Newton's Second Law of Motion.

1.4 Systems of Particles

In this section we will extend the methods and concepts which we have developed for a single particle to a system of particles. The particular system of particles we will have in mind is a rigid body, in which the relative distances between the various particles of the body are always the same at any time, but the discussion will apply to any body which can be considered as composed of point particles. Suppose that there is a particle of mass m_j at the point $\mathbf{r}_j$, where $1 \leqslant j \leqslant N$, there being N particles altogether. The momentum of the jth particle is $\mathbf{p}_j = m_j \dot{\mathbf{r}}_j$, and Newton's Second Law of Motion applied to this jth particle is

$$\mathbf{F}_j = m_j \ddot{\mathbf{r}}_j = \dot{\mathbf{p}}_j \qquad (1.43)$$

where $\mathbf{F}_j$ is the force acting on the jth particle. We define the total momentum $\mathbf{P}$ of the particles to be

$$\mathbf{P} = \sum_{j=1}^{N} \mathbf{p}_j$$

so that using (1.43) for each j we obtain

$$\dot{\mathbf{P}} = \sum_{j=1}^{N} \mathbf{F}_j \qquad (1.44)$$

Now the force on the jth particle has two components, an external force $\mathbf{f}_j$ and a sum of contributions coming from the force $\mathbf{f}_{jk}$ due to the kth particle, for $k \neq j$. Thus

$$\mathbf{F}_j = \mathbf{f}_j + \sum_{k, k \neq j} \mathbf{f}_{jk} \qquad (1.45)$$

From Newton's Third Law of Motion, the force on the jth particle due to the kth particle is equal and opposite to that on the kth particle due to the jth particle, so

$$\mathbf{f}_{jk} = -\mathbf{f}_{kj} \qquad (1.46)$$

If we sum over j on both sides of (1.45), the antisymmetry of $\mathbf{f}_{jk}$ in j and k expressed by (1.46) implies that there is no contribution to this sum over j from the last term on the right-hand side of (1.45):

$$\sum_{j=1}^{N} \mathbf{F}_j = \sum_{j=1}^{N} \mathbf{f}_j = \mathbf{F}_{\text{ext}} \qquad (1.47)$$

where $\mathbf{F}_{\text{ext}}$ is the total external force on the system of particles, being the sum of the external forces $\mathbf{f}_j$ on each particle. Combining (1.44) and (1.47) we obtain

$$\dot{\mathbf{P}} = \mathbf{F}_{\text{ext}} \qquad (1.48)$$

For an isolated system of particles the external force on the system must vanish so that $\mathbf{P}$ = constant. This is the *law of conservation of linear momentum*.

We may also introduce the total angular momentum $\mathbf{L}$ for the system of particles by

$$\mathbf{L} = \sum_{j=1}^{N} \mathbf{L}_j = \sum_{j=1}^{N} \mathbf{r}_j \wedge \mathbf{p}_j = \sum_{j=1}^{N} \mathbf{r}_j \wedge m_j \dot{\mathbf{r}}_j \qquad (1.49)$$

By differentiation of (1.49) we obtain

$$\dot{\mathbf{L}} = \sum_{j=1}^{N} \mathbf{r}_j \wedge m_j \ddot{\mathbf{r}}_j = \sum_{j=1}^{N} \mathbf{r}_j \wedge \mathbf{F}_j \qquad (1.50)$$

If we now use (1.45) and (1.50),

$$\begin{aligned}
\dot{\mathbf{L}} &= \sum_{j=1}^{N} \mathbf{r}_j \wedge (\mathbf{f}_j + \sum_{k \neq j} \mathbf{f}_{jk}) \\
&= \sum_{j=1}^{N} \mathbf{r}_j \wedge \mathbf{f}_j + \sum_{\substack{j,k=1 \\ j \neq k}}^{N} \mathbf{r}_j \wedge \mathbf{f}_{jk} \\
&= \sum_{j=1}^{N} \mathbf{r}_j \wedge \mathbf{f}_j + \sum_{j<k} (\mathbf{r}_j - \mathbf{r}_k) \wedge \mathbf{f}_{jk} \qquad (1.51)
\end{aligned}$$

For a rigid body the force $\mathbf{f}_{jk}$ will be parallel to the position vector joining the particles j and k, which is $(\mathbf{r}_j - \mathbf{r}_k)$. Thus the last term on the right-hand side of (1.51) will vanish, and the first term will just be the external couple, $\mathbf{G}_{\text{ext}}$, so

$$\dot{\mathbf{L}} = \mathbf{G}_{\text{ext}}$$

For an isolated body we have $\mathbf{G}_{\text{ext}} = 0$, so that the total angular momentum will also be a conserved quantity, that is, a constant. It has already been remarked that conserved quantities (or integrals of the motion) are important in determining general features of the motion of a single particle; these quantities are even more helpful when a system of particles is involved, especially since the motion of a large number of particles can be almost impossible to obtain in detail.

We may also develop the Lagrangian and Hamiltonian form of equations of motion equivalent to Newton's Second Law of Motion. For motion in purely conservative fields of force, the total kinetic and potential energies T and V are

$$T = \sum_{j=1}^{N} \mathbf{p}_j^2 / 2m_j, \qquad V = \sum_{j=1}^{N} v_j(\mathbf{r}_j)$$

Defining the Hamiltonian as $H = T + V$, and the Lagrangian as $L = T - V$ (where T is expressed as a function of the velocities $\dot{\mathbf{r}}_j$ by $T = \sum_{j=1}^{N} \frac{1}{2} m_j \dot{\mathbf{r}}_j^2$), the Lagrange equations of motion will be

$$\frac{d}{dt}\left(\frac{\partial L}{\partial \dot{x}_j}\right) - \frac{\partial L}{\partial x_j} = 0 \quad (j = 1, 2, \cdots, N) \qquad (1.52)$$

and similar equations with x_j replaced by y_j or z_j, where

$$\mathbf{r}_j = (x_j, y_j, z_j).$$

The Hamiltonian equations of motion will be

$$\dot{x}_j = \frac{\partial H}{\partial p_{j,x}}, \quad \dot{p}_{j,x} = -\frac{\partial H}{\partial x_j} \quad (j = 1, 2, \cdots, N) \quad (1.53)$$

and similar equations with x_j and $p_{j,x}$ replaced by $y_j, p_{j,y}$ or $z_j, p_{j,z}$, where $p_j = (p_{j,x}, p_{j,y}, p_{j,z})$.

1.5 Variational Principles *

We will finish this discussion of classical mechanics by describing a variational formulation of the above equations. Variational principles have been very important in the development of mechanics, and at some periods have almost played the role of moral principles. The most important quantity is the action A for a given motion of the systems of particles. Suppose this motion is described by specifying the positions and velocities of each of the particles from the time t_1 to time t_2. Then A is defined by

$$A = \int_{t_1}^{t_2} L(\mathbf{r}_1(t), \dot{\mathbf{r}}_1(t); \cdots ; \mathbf{r}_N(t), \dot{\mathbf{r}}_N(t))\, dt \quad (1.54)$$

We consider the change δA in A caused by small changes $\delta \mathbf{r}_j(t)$ in the paths $\mathbf{r}_j(t)$. Then

$$\delta A = \int_{t_1}^{t_2} \{L(\mathbf{r}_1 + \delta\mathbf{r}_1, \dot{\mathbf{r}}_1 + \delta\dot{\mathbf{r}}_1, \cdots, \mathbf{r}_N + \delta\mathbf{r}_N, \dot{\mathbf{r}}_N + \delta\dot{\mathbf{r}}_N)$$
$$- L(\mathbf{r}_1, \dot{\mathbf{r}}_1; \cdots; \mathbf{r}_N, \dot{\mathbf{r}}_N)\}\, dt$$
$$= \int_{t_1}^{t_2} \sum_{j=1}^{N} \{(\partial L/\partial \mathbf{r}_j) \cdot \delta\mathbf{r}_j + (\partial L/\partial \dot{\mathbf{r}}_j) \cdot \delta\dot{\mathbf{r}}_j\}\, dt \quad (1.55)$$

where $\partial L/\partial \mathbf{r}_j$ is the vector with components $(\partial L/\partial x_j, \partial L/\partial y_j, \partial L/\partial z_j)$, and $\partial L/\partial \dot{\mathbf{r}}_j = (\partial L/\partial \dot{x}_j, \partial L/\partial \dot{y}_j, \partial L/\partial \dot{z}_j)$. But $\delta\dot{\mathbf{r}}_j = d/dt\, (\delta\mathbf{r}_j)$, and we write in the last term of (1.55)

$$\left(\frac{\partial L}{\partial \dot{\mathbf{r}}_j}\right) \cdot \left(\frac{d}{dt}\right) \delta\mathbf{r}_j = \frac{d}{dt}\left(\frac{\partial L}{\partial \dot{\mathbf{r}}_j} \cdot \delta\mathbf{r}_j\right) - \delta\mathbf{r}_j \cdot \left\{\frac{d}{dt}\left(\frac{\partial L}{\partial \dot{\mathbf{r}}_j}\right)\right\}$$

Then

$$\delta A = \sum_{j=1}^{N} \delta \mathbf{r}_j \cdot \left(\frac{\partial L}{\partial \dot{\mathbf{r}}_j}\right)\bigg|_{t_1}^{t_2} + \int_{t_1}^{t_2} \sum_{j=1}^{N} \left\{\frac{\partial L}{\partial \mathbf{r}_j} - \frac{d}{dt}\left(\frac{\partial L}{\partial \dot{\mathbf{r}}_j}\right)\right\} \cdot \delta \mathbf{r}_j \, dt \quad (1.56)$$

If we choose the variations $\delta \mathbf{r}_j$ to vanish at the initial and final times t_1 and t_2, then the first term on the right-hand side of (1.56) will vanish. We see that the Lagrange equations of motion (1.52) imply that for small variations from the *actual* paths of motion of the particles then $\delta A = 0$; conversely if we require that $\delta A = 0$ for small variations from the actual paths then the actual paths must satisfy the Lagrange equations (1.52). In this way we may reduce the problems of solving the various equivalent systems of equations of motion to that of finding paths which make the action A stationary for small variations. The variational principle

$$\delta A = 0 \quad (1.57)$$

is called the *principle of least action*.

We may use (1.57) to obtain the Lagrange equations (1.35) for general co-ordinates (and hence the Hamiltonian equations (1.34)). To do this we consider the action (1.54) defined by means of a Lagrangian depending on generalized co-ordinates $q_1, \cdots, q_N$ and the corresponding velocities $\dot{q}_1, \cdots, \dot{q}_N$. If we take small variations $\delta q_1(t), \cdots, \delta q_N(t)$ of these co-ordinates between the times t_1 and t_2 with vanishing variations initially and finally, we may repeat the steps leading to (1.56) but with the r_j's replaced by the q_j's. We will obtain

$$\delta A = \int_{t_1}^{t_2} \sum_{j=1}^{N} \{(\partial L/\partial q_j) - d/dt \, (\partial L/\partial \dot{q}_j)\} \, \delta q_j(t) \, dt \quad (1.58)$$

Hence $\delta A = 0$ for all small variations $\delta q_j(t)$ will require that the coefficient of each such variation $\delta q_j(t)$ under the integral sign in (1.58) must vanish; this gives the Lagrange equations (1.35).

PROBLEMS

1.1 If a particle is describing an orbit in three dimensions under a force directed to a fixed point O show that the orbit must be a plane curve, and the rate of description of area by the radius vector drawn from the fixed point to the particle is constant.

Use polar co-ordinates (r, θ) in the plane of the motion to obtain equation of motion

$$\ddot{r} - r\dot{\theta}^2 = -f$$

$$r^2\dot{\theta} = \text{constant} = h$$

where f is the force, and using $u = 1/r$, derive the equation of motion

$$f = h^2 u^2 \{(d^2u/d\theta^2) + u\}$$

Solve the equation for an inverse square law force, and discuss the nature of the ensuing motion.

1.2 An electron of charge e is moving round a stationary proton of charge $-e$ in a circular orbit of radius r. What is the potential energy of the electron, and what is its total energy?

If the rate of loss of energy of the electron when it has an acceleration $\ddot{r}$ is $-(2e^2/12\pi\epsilon_0 c^3)(\ddot{r})^2$, how long does it take for a classical hydrogen atom of radius $r_0 = 4\pi\epsilon_0 \hbar^2/me^2$ to shrink to half its size, regarding the electron's orbit as circular throughout the shrinkage.

(You may take $\hbar/mc = 4 \times 10^{-13} m$, $e^2/4\pi\epsilon_0 \hbar c = 1/137$.)

FURTHER READING

1. GOLDSTEIN, H., *Classical Mechanics*, Addison Wesley, Cambridge, Mass., 1950; a very complete discussion, especially good on Hamilton's and Lagrange's Equations.
2. BECKER, R. A., *Introduction to Theoretical Mechanics*, McGraw Hill, 1954; a more introductory discussion.

CHAPTER 2

The Development of Wave Mechanics

2.1 The Difficulties of Classical Mechanics

In the previous chapter we described the basis of classical mechanics, whose main assumptions for a given physical system are:

(a) there exist dynamical variables at a given time t, which we can call $q_i(t)$ with $1 \leq i \leq N$ for some integer N. Each of these variables takes a *definite* value at time t, the state of the system being determined at time t by the particular value that each of the variables takes at that time.

(b) The development with time of the physical system is determined by the development with time of the set of variables $q_i(t)$, $1 \leq i \leq N$, and the state of the system at any time is determined by this development and by the values of the variables $q_i(t)$ at some initial time.

The problem of determining the classical motion of a given physical system reduces, then, to finding the dynamical variables q_i and ascertaining their development with time. These two problems were solved for systems of particles by taking the dynamical variables to be the positions of the particles and their velocities, the development with time being given by Newton's Laws of Motion. The initial values of the positions and velocities of the particles then determined their later positions, provided that the forces acting on the particles were known. Suitable force laws were found which enabled phenom-

ena involving distances larger than about 10^{-6} metres to be well understood by classical mechanics. The force laws were
 (i) the inverse square law of gravitational attraction between two massive bodies.
 (ii) Coulomb's inverse square law of repulsion or attraction between two charged bodies, suitably generalized to Maxwell's equations and Lorentz force to describe light and the interaction between light and charged particles.

However, this classical description of the motion of particles and light began to break down at the beginning of this century; certain phenomena involving distance of the order of 10^{-10} metres could not be explained classically, and even contradicted the expected classical result. In order to explain these recalcitrant phenomena a new mechanics had to be developed to replace classical mechanics; this was called quantum mechanics, and was found to explain such phenomena. It also agreed very well with classical mechanics when the latter worked.

Quantum mechanics had two stages of development. The first, beginning with Planck's introduction of the quantum of action in 1901, lasted till about 1925. It consisted of a mixture of classical and non-classical concepts, and was not completely satisfactory. The second and final stage was achieved from different but equivalent viewpoints by Heisenberg and Schrödinger in 1925. The difficulties of the older version of quantum mechanics were completely cleared up by this new quantum mechanics (sometimes called wave mechanics because all particles are replaced by waves in Schrödinger's version of it). We should point out here that wave mechanics itself requires corrections. A more complete theory of particles called quantum field theory has been accepted since 1947, and agrees with wave mechanics where the predictions of wave mechanics have been successful. Fig. 2.1 shows how the different particle theories developed over the years.

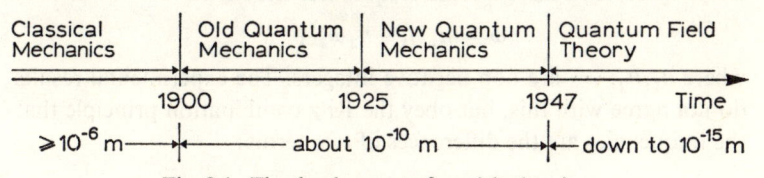

Classical Mechanics	Old Quantum Mechanics	New Quantum Mechanics	Quantum Field Theory
	1900	1925	1947 Time
$\geqslant 10^{-6}$ m	about 10^{-10} m		down to 10^{-15} m

Fig. 2.1 The development of particle theories.

To appreciate more fully the development of wave mechanics, it will be helpful to describe briefly the recalcitrant phenomena (or the critical experiments) which could not be explained by classical mechanics. We will attempt to discuss them according to their historical development. The recalcitrant phenomena are as follows:

1. The Stability of Matter. Matter is made up of molecules. These may consist of charged particles (ions) held together by their Coulomb attraction; an example is sodium chloride, in which the positive sodium ion and the negative chlorine ion attract each other. For a system of charged particles in stable equilibrium, it can be shown that the particles cannot be at rest but must be accelerating. In the process they emit radiation (a charged particle emits light when it accelerates). Such radiation is not seen. What is worse is the motion of the electron round the nucleus; the electron is continuously accelerating towards the nucleus, so should be continuously radiating light. It will lose energy by this means, and rapidly collapse into the nucleus (in a time of the order of 10^{-10} seconds). Such a collapse would be catastrophic for the chemical behaviour of the atoms composing our bodies, and our own continuing existence over times much longer than 10^{-10} seconds indicates that such collapse does not occur. We will not be able to give a simple explanation of the stability of matter till we come to wave mechanics (which, however, was developed by trying to explain other phenomena). We will return to the stability of matter later.

2. Optical Spectra. The light emitted from elements, caused either by heating them or submitting them to an electrical discharge when they are gaseous, is found to have a discrete range of frequencies. On a classical picture these frequencies should be the same as the frequencies of the periodic motions of the charged particles in the atoms, together with their harmonics. If $\omega_1, \omega_2, \cdots$ are these allowed frequencies, then the observed frequencies should be

$$\omega = n_1\omega_1 + n_2\omega_2 + \cdots$$

where $n_1, n_2, \cdots$ are non-negative integers. The experimental results do not agree with this, but obey the Ritz combination principle that the frequencies are the differences of two terms:

$$\omega = \omega'_n - \omega'_m$$

where $\omega'_1, \omega'_2, \cdots$ are a fixed set of frequencies. This is another phenomenon which we will not be able to explain immediately, but will return to it later.

3. Black body Radiation and the Quantum.
We next consider electromagnetic radiation in an enclosure in equilibrium with its surroundings. The radiation is called black body radiation, since if a small hole is made in the enclosure, to allow radiation to escape and so be observed, radiation entering the enclosure through the hole will have almost no chance of escaping; the small hole thus acts as a perfect absorber of radiation, and so appears as black.

If we suppose that the radiation is inside a square box of side L, then each allowed frequency occurs in a plane wave of the form $e^{i\mathbf{k}\cdot\mathbf{r}}$, where $\mathbf{k}\cdot\mathbf{r} = k_x x + k_y y + k_z z$. In order that the wave remain in the box, being reflected back and forth from the walls, it is necessary that it have the same value at one wall as at the opposite one in the box (in equilibrium the wave will not be able to distinguish the opposite walls) so that the allowed values of $\mathbf{k}$ will be $\mathbf{k} = (2\pi/L)(n_1, n_2, n_3)$, where n_1, n_2, n_3 are integers. We want to calculate the number of the allowed values of $\mathbf{k}$; this will give us the number of degrees of freedom for the radiation in the enclosure. We may regard the vectors $\mathbf{n}$ as points in a lattice; the number in a shell of thickness Δn and radius n about the origin will be equal to the volume of the shell (when n and Δn are not too small), that is $\mathbf{n} = 4\pi n^2 \Delta n$ where $n^2 = n_1^2 + n_2^2 + n_3^2$. Thus the number of values of $\mathbf{k}$, with $|\mathbf{k}|$ lying between k and $k+dk$, will be $2(L/2\pi)^3 4\pi k^2 dk$ (The extra 2 is due to the two possible polarization states of an electromagnetic wave.) Since the wave frequency ω is equal to $kc/2\pi$, then the number $n(\omega)d\omega$ of waves with frequency between ω and $\omega + d\omega$ per unit volume of enclosure is

$$n(\omega)\,d\omega = \frac{1}{L^3}\cdot 2\left(\frac{L}{2\pi}\right)^3 4\pi\omega^2 d\omega \left(\frac{2\pi}{c}\right)^3 = \frac{8\pi\omega^2\,d\omega}{c^3}$$

Since the principle of equipartition of energy indicates that each wave has energy kT, where k is Boltzmann's constant and T the temperature of the enclosure, the energy per unit volume of waves with frequency between ω and $\omega + d\omega$ is

$$E(\omega)\,d\omega = \frac{8\pi\omega^2\,d\omega}{c^3}\cdot kT$$

This is the Rayleigh-Jeans formula. It fits the experimental data well for low frequencies, but not for higher ones. This breakdown is to be expected, since the total energy of radiation per unit volume of enclosure is $\int_0^\infty E(\omega)\,d\omega$, which would be infinite! This is evidently not the case, and indeed the total energy can be measured, so must be finite.

This paradox was resolved by Planck in 1900 by a very radical proposal which was the start of the quantum theory. He suggested that radiation of a given frequency ω, say, can only exchange energy with matter in discrete packets or quanta, each of energy $h\omega$. The constant of proportionality h is now called Planck's constant; it has the dimensions of energy times time. The only available energies for the wave of frequency ω will be $0, h\omega, 2h\omega, \cdots$; the probability of having energy $nh\omega$ at temperature T will be $e^{-nh\omega/kT}$, so that the mean energy in the wave of frequency ω will be

$$h\omega \sum_{n \geqslant 0} n e^{-nh\omega/kT} \Big/ \sum_{n \geqslant 0} e^{-nh\omega/kT} = h\omega/(e^{h\omega/kT}-1)$$

Then

$$E(\omega) = (8\pi h \omega^3/c^3)/(e^{h\omega/kT}-1)$$

which is known as Planck's law. This will give a finite total energy of radiation per unit volume of enclosure; it also agrees with the Rayleigh-Jeans formula for small ω. The shape of $E(\omega)$ is shown in Fig. 2.2, and compared with the Rayleigh-Jeans formula. The total emitted energy is

$$E = \int_0^\infty E(\omega)\,d\omega = (4\sigma/c)T^4$$

where $\sigma = (2\pi^5 k^4/15 h^3 c^2)$, and is known as Stefan's constant. Since the values of σ, k, and c are known, we may determine h, giving a value of $h = 6 \cdot 625 \times 10^{-34}$ joule seconds.

Since energy times time has the dimensions of action (as we saw in 1.54, Chap. 1), h is called the constant of action. It is the constant which indicates the degree of discreteness which must be present to explain the energy in black body radiation. This discreteness is absolutely foreign to classical mechanics; it is the basic property of quantum mechanics.

4. The Photoelectric Effect and the Photon. When ultra-violet light falls on a metal surface it is found that a current of electrons flows,

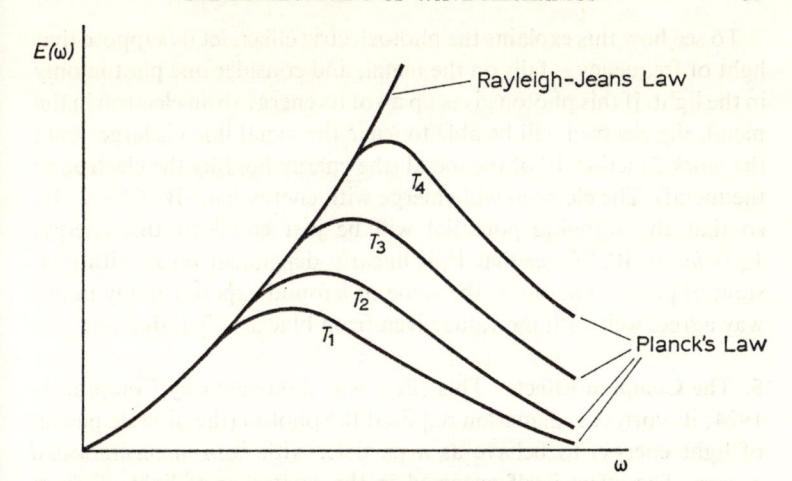

Fig. 2.2 Comparison between the Rayleigh-Jeans and Planck formulae for the energy distribution $E(\omega)$ in black body radiation, as a function of frequency ω. The different temperatures T_1, T_2, T_3, T_4 satisfy $T_1 < T_2 < T_3 < T_4$. The experimental values for $E(\omega)$, at a given temperature, lie very close to curves given by Planck's formula.

even when a retarding potential is present which is not too big. For a large enough retarding potential no electrons are emitted. The largest retarding potential still allowing electrons to be emitted is called the stopping potential V_s, and is proportional to the maximum energy of the electrons emitted from the metal. On a classical picture, the energy of the ultra-violet light will increase with its intensity. The electrons are gaining energy from this light, so that their energy will increase with the light intensity. Thus we expect that if the light intensity is increased the stopping potential should increase also. It was found experimentally that the stopping potential was independent of the light intensity, but increased linearly with the frequency of the incoming light. This certainly could not be explained on the basis of classical mechanics; indeed changing the frequency should have no effect whatsoever on the stopping potential.

The logical extension of Planck's quantum hypothesis was made by A. Einstein in 1905 to explain the photoelectric effect. He suggested that not only is energy exchanged between matter and radiation in discrete packets, but also that radiation actually consists of discrete packets, called *photons*, each of energy $\hbar\omega$ (for light of frequency ω). Thus light would be corpuscular.

To see how this explains the photoelectric effect, let us suppose that light of frequency ω falls on the metal, and consider one photon only in the light. If this photon gives up all of its energy to an electron in the metal, the electron will be able to leave the metal if $h\omega$ is larger than the work function W of the metal (the energy holding the electron in the metal). The electron will emerge with energy $h\omega - W$ if $h\omega > W$, so that the stopping potential will be just equal to this energy; $V_s = h\omega - W$. We see that V_s is linearly dependent on ω with constant of proportionality h; the value of h found experimentally in this way agrees well with the value given from black body radiation.

5. The Compton Effect.

This effect was discovered by Compton in 1924; its correct explanation required the photon (the discrete packet of light energy) to behave as a particle, with *both momentum and energy*. The effect itself occurred in the scattering of light of short wave length (X-rays with wavelength of about 10^{-10} metres, or about the size of an atom) by free or weakly bound electrons. It was found

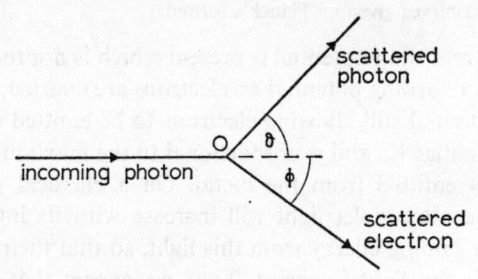

Fig. 2.3 Picture of the Compton scattering process showing a photon scattered from an electron which is initially at rest at O.

that the change in wave length $\Delta\lambda$ in the scattering process was related to the angle through which the X-rays had been scattered by the Compton formula:

$$\Delta\lambda = (2h/mc)(\sin\theta/2)^2$$

where m is the mass of the electron and c the velocity of light. This formula is not in agreement with that expected by classical mechanics; indeed classically the electron will absorb the incoming radiation. This causes it to accelerate, so that it will subsequently radiate, and it may be shown that the difference $\Delta\lambda$ between the wavelength of the incoming and emitted radiation is proportional to λ. This evidently disagrees with the Compton formula.

The correct explanation is that the incident light is composed of a beam of photons, each moving with velocity c, energy $h\omega$, and *momentum $h\omega/c$*. We now impose energy and momentum conservation and let ω and ω' be the initial and final frequencies of the radiation, and p the final momentum of the electron at an angle ϕ to the direction of the incoming radiation, as is shown in Fig. 2.3. The energy and momentum of the initial and final photons are $(h\omega, h\omega/c)$ and $(h\omega', h\omega'/c)$ whilst that of the initial and final electrons are $(mc^2, 0)$ and $((c^2p^2+m^2c^4)^{\frac{1}{2}}, p)$, with momentum directions as shown in Fig. 2.3. Then the conservation of energy requires

$$mc^2 + h\omega = (m^2c^4 + p^2c^2)^{\frac{1}{2}} + h\omega'$$

and conservation of momentum along the initial direction of the radiation requires

$$h\omega/c = (h\omega'/c)\cos\theta + p\cos\phi$$

whilst that perpendicular to this initial direction requires

$$(h\omega'/c)\sin\theta + p\sin\phi = 0$$

Hence

$$h\{(\omega - \omega'\cos\theta)^2/c^2\} + \{(h\omega'\sin\theta)/c\}^2 = p^2$$

or

$$(h^2/c^2)(\omega^2 - 2\omega\omega'\cos\theta + \omega'^2) = p^2$$

But

$$p^2 = \{mc + (\omega - \omega')(h/c)\}^2 - m^2c^2$$

and also

$$c = \omega\lambda = \omega'\lambda'$$

so that

$$\Delta\lambda = \lambda' - \lambda = c\left(\frac{1}{\omega'} - \frac{1}{\omega}\right) = \frac{c(\omega - \omega')}{\omega\omega'}$$

Hence

$$\frac{h^2}{c^2}\{(\omega - \omega')^2 + 4\omega\omega'\sin^2\tfrac{1}{2}\theta\} = 2mh(\omega - \omega') + \frac{h^2}{c^2}(\omega - \omega')^2$$

so

$$2mh(\omega - \omega') = 4\omega\omega'h^2\sin^2\tfrac{1}{2}\theta/c^2$$

or
$$\Delta\lambda = (2h/mc)(\sin\tfrac{1}{2}\theta)^2$$
which is Compton's formula.

We have dealt in this section with the concept that light is composed of discrete particles called photons, each of which behaves as a particle of energy $h\omega$ and momentum $h\omega/c$; in other words light is quantised. We now turn to matter to see how it should be quantised in order to explain the stability properties of matter and the optical spectra of atoms.

2.2 The Quantisation of Matter

The problems of the stability of matter and the optical spectra of atoms still arose for the Rutherford model of the atom. This model involved a central massive positively charged nucleus, around which the electrons moved under the Coulomb attraction between them and the nucleus. The first step in the quantisation of matter was made by Niels Bohr in 1913. He set up rules which solved both the problem of stability and that of optical spectra. These rules were:

(i) There exists a discrete set of stationary (or stable) states of an atom; from each of these states no electromagnetic radiation is emitted due to the accelerating motion of the electrons round the nucleus.
(ii) Transitions occur from one of these states to another with the emission of radiation. If the initial and final states have energies E_n and E_m the frequency of the emitted radiation is $(E_n - E_m)/h$; this agrees with the Ritz combination principle which we discussed earlier in 2.1 of this chapter.
(iii) The angular momentum of an electron in motion round the nucleus in a hydrogen atom takes the values $0, \hbar, 2\hbar, 3\hbar, \cdots$ only, where $\hbar = h/2\pi$. This quantisation of angular momentum is very basic in quantum mechanics.

Let us consider the implications of the above three rules for the hydrogen atom, and let us take the electron to be moving in a circular orbit of radius r and with velocity v. Let e be the charge on the proton and $-e$ that on the electron. We have that the angular momentum of the electron about the nucleus is $J = mvr$, which is equal to $n\hbar$ for some integer n, by rule (iii). Also the electron is under an acceleration v^2/r towards the nucleus and a force $e^2/4\pi\epsilon_0 r^2$, so for stability $mv^2/r = e^2/4\pi\epsilon_0 r^2$. Then

$$r = 4\pi\epsilon_0 n^2\hbar^2/me^2 = a_n$$

where $a_1 = 4\pi\hbar^2\epsilon_0/me^2$ is called the Bohr radius of the atom, and is equal in size to 0.53×10^{-10} metres. The allowed energy levels of the electron are the values

$$E_n = \text{kinetic energy} + \text{potential energy}$$
$$= \tfrac{1}{2}mv^2 - (e^2/4\pi\epsilon_0 r) = -me^4/2(4\pi\epsilon_0 n\hbar)^2.$$

The energy levels are discrete, and by rule (ii) above the frequency ω of radiation emitted in a transition from level n_1 to level n_2 (with $n_1 > n_2$) is

$$\omega = \frac{2\pi^2 me^4}{ch^3(4\pi\epsilon_0)^2}\left(\frac{1}{n_2^2} - \frac{1}{n_1^2}\right)$$

This formula fits the frequencies of the lines in the optical spectrum of hydrogen. These had been previously classified in series, the Lyman, Balmer, Paschen, etc., series; each of these series was found to correspond to transitions from various energy levels down to a particular value of n_2, with $n_2 = 1$ for the Lyman series, $n_2 = 2$ for the Balmer series, and so on.

To calculate the energy level E_1 we use that the rest energy of the electron is about $\tfrac{1}{2} MeV$ (1 MeV = 1 million electron volts) and the dimensionless coupling constant for the Coulomb interaction is $\alpha = e^2/4\pi\epsilon_0 \hbar c = 1/137$ (α is called the fine structure constant). In terms of these two quantities

$$E_1 = -\tfrac{1}{2}mc^2\left(\frac{e^2}{4\pi\epsilon_0 \hbar c}\right)^2 = -\left(\frac{1}{4}\right)\left(\frac{1}{137}\right)^2 MeV = -13\cdot 6\, eV.$$

The higher energy levels are given by $E_n = -(13\cdot 6/n^2)\, eV$. To calculate the frequencies of lines emitted in a transition from the level n_1 to the level n_2, we have to use the values of h. In terms of electron volts as energy units $h/2\pi = 6\cdot 6 \times 10^{-16}\, eV$ seconds. Then the frequency, ω, of light emitted when an electron drops from energy level E_2 to level E_1 is $\omega = 2.5 \times 10^{15}$ s^{-1}, whilst the wave length of this radiation is $\lambda = c/\omega = 1.2 \times 10^{-7}$ m, where $c = 3 \times 10^8$ m s^{-1}. Such light is in the ultra-violet region, the visible region being from 4×10^{-7} to 8×10^{-7} m.

A schematic diagram of the energy levels and the transitions in the hydrogen atom are shown in Fig. 2.4. We see that all the energy levels are negative if we take the energy E_∞ of an electron at rest at an infinite distance from the proton to be zero. Then the energy required to remove an electron from the lowest energy E_1, called the ionization energy, will be $-E_1$.

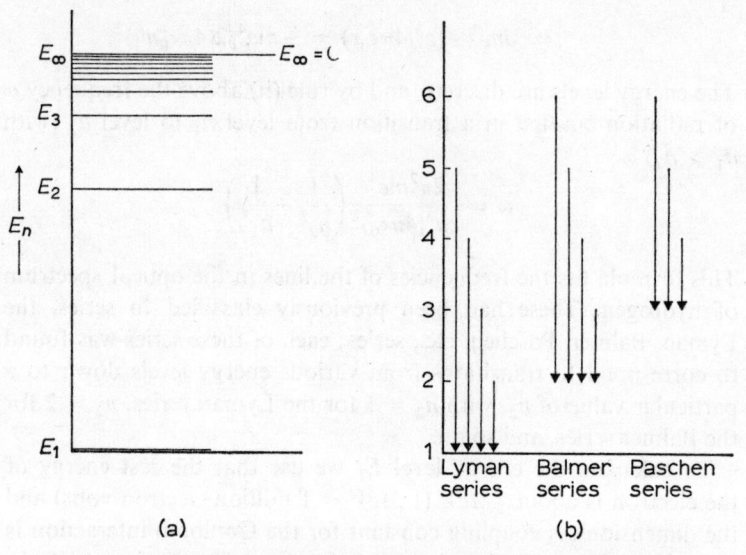

Fig. 2.4 (a) Schematic diagram of the energy levels in the hydrogen atom. E_1 is the lowest energy level, E_∞ is taken to be zero, so that E_n is negative for all n, and $E_1 < E_2 < E_3 < \cdots < 0$.
(b) Transitions in the hydrogen atom giving spectral lines in the various series. Each vertical line corresponds to a transition from the energy level with quantum number n_1, at the upper end of the line to the level with quantum number n_2 at the lower end of the line.

Rule (iii) above was extended to periodic but non-circular orbital motion of the electron by *the Sommerfeld-Wilson rule*:

For any periodic motion described by a set of co-ordinates $q_1, \cdots, q_N$ and corresponding moment $p_1, \cdots, p_N$ the N phase integrals $J_i = \oint p_i dq_i$ $((i = 1, 2, \cdots, N)$ take the values $J_i = n_i h$ (n_i an integer), where $\oint$ denotes integration over one period of the motion.

This rule evidently extends rule (iii) for circular motion of the electron to the more general case of elliptic orbits; in the circular case we take $N = 1, q_1 = \theta$, where θ is the angle describing the motion of the electron round the circle, and $p_1 = p_\theta$, which is the constant angular momentum mvr, and $\oint$ denotes an integral once round the circle, so is from $\theta = 0$ to $\theta = 2\pi$. Thus $J = 2\pi mvr$, and the rule tells us $2\pi mvr = nh$, so $mvr = n\hbar$, which is rule (iii).

It is straightforward to show that the elliptical orbits have the same possible set of energies as the circular orbits. We won't give that discussion here since we will have a more complete discussion of the hydrogen atom by means of the Schrödinger equation. However, we will use the Sommerfeld-Wilson rule to quantise a particle moving in a box with perfectly reflecting walls. Whilst this system doesn't appear very important it will lead us to an important principle which enables us to relate classical and quantum systems.

Consider a particle moving freely in a box in one dimension, and let a be the width of the box. If p is the momentum of the particle in the box, then the particle oscillates back and forth with this momentum. Thus $\oint p_x dx$ will have two contributions, one from the part of the motion when the particle moves to the left, and the other when it moves to the right, as shown in Fig. 2.5. When the particle is mov-

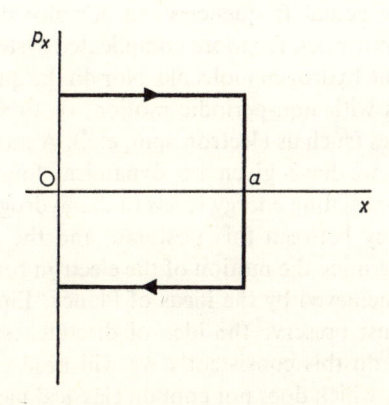

Fig. 2.5 The phase plane diagram of (x, p_x) for the motion of a particle moving freely in a one-dimensional box with perfectly reflecting walls. The arrowed path shows the actual phase plane motion.

ing to the right, $p_x = p$, whilst when it is moving to the left, $p_x = -p$, but also dx has changed sign. Thus $J = \oint p_x \, dx = 2pa$ and by the Sommerfeld-Wilson rule $J = nh$, so the possible energies are

$$E_n = \frac{p^2}{2m} = \frac{n^2 h^2}{8ma^2} \quad (n = 0, 1, 2, \ldots).$$

For a particle of mass one gram in a box of side $10^{-2} m$ we can calculate that the difference $E_2 - E_1 = 10^{-48} \, eV$. In other words, for a macroscopic body the discreteness of the energy levels is negligible. Thus we have *the correspondence principle*:

For a macroscopic system the discrete energy levels become so close that the system is indistinguishable from one with a continuous range of energies i.e. a classical system.

We will return to this principle later; it will prove important in determining the form of the quantum equations of motion.

2.3 Matter Waves

The quantisation rules of the last section enabled the energy levels of the hydrogen atom to be obtained. Whilst this was an important advance, numerous difficulties were still left unsolved, e.g. the relative intensities of the spectral lines cannot be predicted for the hydrogen atom, though the actual frequencies can be; also the Sommerfeld-Wilson rules do not work for more complicated systems, such as the helium atom or the hydrogen molecule. Nor do the quantisation rules apply to systems with non-periodic motion, or to systems without classical analogues (such as electron spin, etc.). A more fundamental difficulty is that we have given no dynamical foundation for the postulate of non-radiating energy levels in the hydrogen atom. There is an inconsistency between this postulate and the use of classical mechanics to determine the motion of the electron round the proton.

The successes achieved by the ideas of Planck, Einstein and Bohr show that we must preserve the idea of discreteness and of stable energy levels; to do this consistently we will need to set up a new dynamical theory which does not contain classical mechanics. At the same time we must be able to explain, using our new theory, the successes of classical mechanics in the macroscopic world.

The crucial fact necessary to set up such a new dynamical theory is that matter exhibits *wave-like properties*. That this is so was first predicted by de Broglie in 1924, and was demonstrated experimentally by Davisson and Germer in 1927, who obtained a diffraction pattern by passing an electron beam through a crystal grating. Indeed such a property is to be expected if we argue by analogy: light is a wave, since it exhibits well known interference behaviour under certain conditions; we saw above that it also behaves as a set of particles (the photon) under certain conditions. Matter evidently has particle properties, so by analogy with light it should also have wave properties under suitable conditions. The question is: what are these conditions? The answer to this question is very important; when the conditions are valid we do not expect classical mechanics to apply, and evidently we have to develop a new mechanics, which it is natural to call wave mechanics. Alternatively, when the matter being investigated behaves as a particle or set of particles, we might expect classical mechanics to apply.

The wave properties of matter are partly determined by the wave length. This can be measured, for example, by setting up a diffraction pattern when a beam of electrons passes through a nickel crystal. Consider the diffraction of a beam of wave length λ at right angles to the atoms in a particular plane, where the atoms are separated from each other by a distance d (see Fig. 2.6). The beam deflected through an angle θ will have maximum intensity if the distance BC, the difference between two adjacent paths, is an integral number of wave lengths. In other words maxima occur for

$$BC = d \sin \theta = n\lambda \tag{2.1}$$

These peaks are the Bragg diffraction peaks; if d is known (by using beams of known wave length, say X-rays) then we may determine λ from the position of the first maximum ($n = 1$). It was found that the relation between the wave length λ and the momentum p of the electrons is

$$\lambda = h/p \tag{2.2}$$

In other words the wave length of the electron beam is inversely proportional to the momentum of the beam. By using thin metal films, similar diffraction patterns were obtained by G. P. Thomson for beams composed of other particles, such as beams of positively

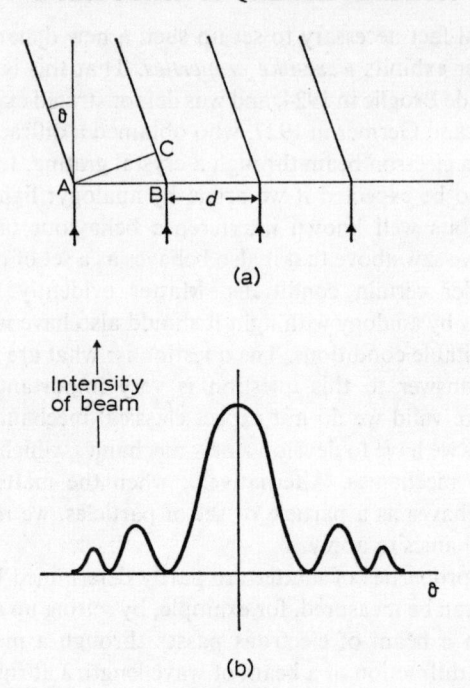

Fig. 2.6 (a) The formation of a diffraction pattern in light falling onto the atoms in a crystal. The path difference BC between two adjacent paths determines whether the rays travelling along these paths after being scattered from the atoms at A and B will interfere constructively or destructively.
(b) A typical diffraction pattern, showing the various maxima and minima arising from constructive or destructive interference.

charged nuclei (protons, O^{16} nuclei, etc.). Indeed, all other particles are found to have a wave nature with wave length given by (2.2); this relation is known as *the de Broglie relation*. Typical values of λ are shown in Table 1. We expect effects connected with Planck's constant of action h to be small when the masses and energies of particles involved over periods of a second or more are much larger than h; in such situations we may take $h = 0$, so that $\lambda = 0$. Thus the classical limit is expected to arise when the wave length of the particles being studied is very small with respect to other lengths in the situation. Otherwise we expect the wave nature of matter to be important in determining the behaviour of the system being studied.

This wave/particle duality may be analysed further by considering how we may build up a localized object from waves. We may do the same as we did for light, i.e. take a wave packet—a linear superposition of plane waves—chosen so that they interfere in a constructive or additive fashion in a localized region of space, and interfere destructively elsewhere. If the localization is maintained in the course of time, then we can speak of it as a particle. This will be like a point particle if there are many waves of very short wave length, which corresponds to the result we obtained above, that the classical limit is obtained for very short wave lengths.

TABLE 1

Wave lengths as given by the de Broglie formula (2.2)

Particle Beam	Kinetic Energy (MeV)	Wavelength (Metres)
Electrons	10^{-6}	$1 \cdot 2 \times 10^{-9}$
	1	$8 \cdot 2 \times 10^{-13}$
Protons	10^{-6}	2×10^{-11}
	1	2×10^{-14}
Oxygen Nuclei	10^{-6}	5×10^{-12}
	1	5×10^{-15}

To make the idea of a wave packet concrete, let us consider its mathematical expression. A plane wave is represented by the wave function

$$e^{i(\mathbf{k} \cdot \mathbf{x} - \omega t)}$$

This wave has frequency $\nu = \omega/2\pi$, wave length $\lambda = 2\pi/k$, ($k = |\mathbf{k}|$), and velocity $v = \omega/k$. We now form a wave packet

$$\psi(\mathbf{x}, t) = \int f(\mathbf{k}) e^{i(\mathbf{k} \cdot \mathbf{x} - \omega t)} d^3\mathbf{k} \qquad (2.3)$$

In the expression (2.3) we will assume that ω is a given function of $\mathbf{k}$, $\omega = \omega(\mathbf{k})$, so that in general there will be dispersion in the wave (2.3), the wave packet spreading out as it moves along. To reduce this dispersion as much as possible we take $f(\mathbf{k})$ to be non-zero only over a small interval,

$$(\mathbf{k}_0 - \Delta \mathbf{k})_i \leqslant k_i \leqslant (\mathbf{k}_0 + \Delta \mathbf{k})_i$$

with $\Delta k_i \ll k_{0i}$ and $i = 1, 2, 3$. Then we may take

$$\omega(\mathbf{k}) = \omega_0 + \sum_{i=1}^{3} (k_i - k_{0i})(\partial\omega/\partial k_i)_{\mathbf{k}_0} = \omega_0 + (\mathbf{k} - \mathbf{k}_0) \cdot \mathbf{v}_g$$

where $(\mathbf{v}_g)_i$ has been defined as $(\partial\omega/\partial k_i)_{\mathbf{k}_0}$. The wave function $\psi(\mathbf{x}, t)$ of (2.3) becomes

$$\int_{\mathbf{k}_0 - \Delta \mathbf{k}}^{\mathbf{k}_0 + \Delta \mathbf{k}} d^3\mathbf{k} \; e^{i(\mathbf{k}_0 \cdot \mathbf{x} - \omega t)} e^{i(\mathbf{x} - \mathbf{v}_g t) \cdot (\mathbf{k} - \mathbf{k}_0)} \tag{2.4}$$

The wave packet of (2.4) is composed of two factors: the first factor represents a wave of frequency $\omega_0/2\pi$ and wave length $2\pi/k_0$; the last factor describes a modulation of the amplitude of this wave. This modulation moves with velocity $\mathbf{v}_g$, so that (2.4) describes a group of waves moving with group velocity $\mathbf{v}_g$.

Let us suppose that the energy E and frequency ν of a plane wave are related by the Planck frequency condition $E = h\nu$; this was true for electromagnetic waves, so that our analogy between the dual properties of electromagnetic radiation and matter allows us to assume it for matter also. Thus $E = \hbar\omega$, so that the group velocity of a wave packet of mean wave length $2\pi/k_0$ has components $(\partial\omega/\partial k_i)_{\mathbf{k}_0} = (\partial E/\partial k_i)_{\mathbf{k}_0}/\hbar$. If we require that this group velocity is equal to the velocity $\mathbf{v}$ of the particle represented by the wave packet, where $\mathbf{v} = (\partial E/\partial p_1, \partial E/\partial p_2, \partial E/\partial p_3)$ with $\mathbf{p}$ the particle momentum, then $\partial E/\partial p_i = \partial E/\partial(\hbar k_{0i})$ ($i = 1, 2, 3$). Thus $\mathbf{p} = \hbar\mathbf{k}_0$, which, when expressed in terms of the wavelength $\lambda = 2\pi/k_0$ gives $\lambda = h/p$, the de Broglie relation (2.2). Since this relation is experimentally correct, we may suppose our assumptions that there is a close analogy between matter and light and that a wave packet describes a particle, to be valid. We will use these assumptions to set up the dynamical equation describing how matter waves develop in time.

2.4 The Schrödinger Equation

Let us consider a single particle. We saw in the last section that we can construct a particle by means of a localized wave packet. We will now consider the development with time of such a wave packet. Let us restrict ourselves initially to a freely moving particle. If the particle momentum is p and its mass is m then the particle's energy is $E = p^2/2m$. We want to describe the particle by a localized wave

THE DEVELOPMENT OF WAVE MECHANICS

packet of the form of (2.3); such a description does not necessarily restrict the form of the function $f(\mathbf{k})$ or the dependence of ω on k. However, if we require that each of the constituent plane waves $e^{i(\mathbf{k}\cdot\mathbf{x}-\omega t)}$ has the same energy-momentum relationship as for the whole wave packet (2.3) describing the particle, then the energy $\hbar\omega$ is equal to $\hbar^2 k^2/2m$ for such a plane wave. Thus we have the dispersion formula

$$\hbar\omega = \hbar^2 k^2/2m \qquad (2.5)$$

for the wave packet (2.3). We may now obtain a differential equation describing the development with time of the wave packet, for we evaluate

$$-(\hbar^2 \nabla^2/2m)\psi(\mathbf{x}, t) = \int (\hbar^2 k^2/2m) f(\mathbf{k}) e^{i(\mathbf{k}\cdot\mathbf{x}-\omega t)} d^3\mathbf{k}$$

By the dispersion formula (2.5) this is equal to

$$\int \hbar\omega f(\mathbf{k}) e^{i(\mathbf{k}\cdot\mathbf{x}-\omega t)} d^3\mathbf{k} = i\hbar \partial\psi/\partial t$$

Thus we obtain the *Schrödinger equation* for a free particle:

$$i\hbar\, \partial\psi/\partial t = (-\hbar^2 \nabla^2/2m)\psi \qquad (2.6)$$

We may rewrite this equation in a more transparent and useful form if we recognize that the operator $(-\hbar^2 \nabla^2/2m)$ is obtained from the Hamiltonian for the free particle, $H(\mathbf{p}) = \mathbf{p}^2/2m$, by the replacement

$$\mathbf{p} \to (\hbar/i)\nabla \qquad (2.7)$$

$$H(\mathbf{p}) \to H((\hbar/i)\nabla) \qquad (2.8)$$

and the Schrödinger equation (2.4) is obtained by the further replacement

$$E \to i\hbar\, \partial/\partial t \qquad (2.9)$$

If the replacement (2.8) and (2.9) are made in the equation

$$E\psi = H\psi \qquad (2.10)$$

then we obtain exactly the Schrödinger equation (2.6).

We may now extend the Schrödinger equation to the case of a particle moving in a potential so that now it has the Hamiltonian $H(\mathbf{r}, \mathbf{p}) = (\mathbf{p}^2/2m) + V(\mathbf{r})$. We may extend (2.10) with this Hamiltonian, via the substitution (2.7) in the form

$$H(\mathbf{r}, \mathbf{p}) \to H(\mathbf{r}, (\hbar/i)\nabla) \qquad (2.8')$$

and (2.9) to give

$$i\hbar\, \partial\psi/\partial t = [(-\hbar^2 \nabla^2/2m) + V(\mathbf{r})]\psi \qquad (2.6')$$

The Schrödinger equation (2.6') is the basic equation of wave mechanics; it specifies how the wave packet describing the particle develops in the course of time while interacting with a given potential field.

We have now derived the Schrödinger equation from two basic ideas. The first is that of the analogy of the wave properties of matter and those of light; the de Broglie relationship (2.2) derived in the previous section on this basis was found to be experimentally correct, and so gives support for the use of this analogy. The second idea is embodied in the dispersion formula (2.5), which we saw corresponds to the substitution formula (2.8). We have as yet no justification for this formula or its extension to (2.8'). In fact the proof of the pudding is in the eating; we will see that the Schrödinger equation which followed from the substitution formula (2.8) or (2.8') solves the difficulties mentioned at the beginning of the previous section, *viz*, the quantisation of aperiodic systems, the determination of the intensities of spectral lines, and above all the reason for the existence of stationary, non-radiating states for electrons in atoms. Indeed, the purpose of the rest of this book is to work out in some detail the way these (and other) difficulties are solved by means of the Schrödinger equation.

To proceed, we need to clarify what we mean by the phrase 'a particle is represented by the localized wave packet ψ'. To do this we must specify precisely how quantities such as the position, momentum, etc., of the particle are obtained from its wave function $\psi(\mathbf{r}, t)$. We see that the Schrödinger equation is a linear equation in ψ, so that if ψ_1, ψ_2 are two wave functions satisfying the Schrödinger equation, any linear combination $(a\psi_1 + b\psi_2)$ will also satisfy it. In other words the wave functions describing a particle satisfy the *principle of superposition*. Now if the particle is described by the wave function ψ_1 we expect it to be moving in a certain fashion, so we may say that ψ_1 describes a particular *state* of the particle. Similarly ψ_2 describes another state of the particle. We now ask, what is to distinguish these two states from one another, or from their linear superposition $(a\psi_1 + b\psi_2)$? A distinction can be obtained if the observables for the particle, its position, etc., are determined in a non-linear fashion from the wave function. Such a non-linear distinction is given by the *probability interpretation of* ψ: the relative probability of finding the particle in the volume dV around the point $\mathbf{r}$ at time t is $|\psi(\mathbf{r},t)|^2\,dV$.

We have stated this interpretation of ψ without any protective coating to make it more digestible. We gave such a coating to the Schrödinger equation (2.6), (2.6') partly to show the continuity and naturalness of this equation as arising from classical mechanics and the wave theory of light. It is evident that we cannot give such a discussion for the probability interpretation of ψ; it is a completely different world from the classical world in which particles are at particular points with *absolute certainty*. The probability interpretation destroys this certainty with a vengeance; indeed for a plane wave $e^{i(\mathbf{k}\cdot\mathbf{x}-\omega t)}$ the particle described by it would have relative probability dV of being in the volume dV for *any point* $\mathbf{r}$, so it is uniformly spread out through space. Thus particles are now spread out in general. The probabilistic nature of the world which we have introduced has caused a great deal of philosophical discussion, and has even been rejected by scientists for philosophical reasons; Einstein said 'God cannot play dice with the world'. We will not concern ourselves with that here, but see that the probability interpretation, together with the Schrödinger equation, gives a remarkably accurate picture of the world down to 10^{-15} metres.

Let us investigate the probability interpretation further. We must first realize that it means that given a large number of identical copies of the system composed of the particle moving in the potential field, each with wave function ψ, the relative number of times in which the particle is in the volume dV at the point $\mathbf{r}$ when each of these copies is looked at is $|\psi(\mathbf{r},t)|^2\,dV/\int|\psi(\mathbf{r},t)|^2\,dV$. The average value of a function of the particle position, $f(\mathbf{r})$, say, when measured in these identical copies of the system, will then be

$$\langle f \rangle = \int f(\mathbf{r})|\psi(\mathbf{r},t)|^2\,dV \Big/ \int |\psi(\mathbf{r},t)|^2\,dV \qquad (2.11)$$

For a function of position and momentum, $f(\mathbf{r},\mathbf{p})$, say, we extend (2.9) in a natural fashion to:

$$\langle f \rangle = \int \psi^*(\mathbf{r},t) f(\mathbf{r},(\hbar/i)\nabla) \psi(\mathbf{r},t)\,dV \Big/ \int |\psi(\mathbf{r},t)|^2\,dV \qquad (2.12)$$

There are obvious difficulties as to how to define the function $f(\mathbf{r},(\hbar/i)\nabla)$ since the ordering of $\mathbf{r}$ and $(\hbar/i)\nabla$ will matter; thus $x(\hbar/i)\partial/\partial x \neq (\hbar/i)(\partial/\partial x)x$. We will never discuss such ambiguous cases, but only consider functions f which are a sum of a function of the position vector $\mathbf{r}$ only and a function of the momentum operator $(\hbar/i)\nabla$ only.

In order for (2.11) and (2.12) to make sense it is necessary that $\int |\psi(\mathbf{r},t)|^2 \, dV < \infty$. This is not true for a plane wave. In fact we need never consider plane waves; the particles involved in any physical situation are always localized in some suitably large region, so can be described by wave packets. It is sometimes a useful fiction to regard particles as being described by plane waves, but the results obtained by regarding particles in this way must agree with what would have been obtained from (2.11) and (2.12) using localized wave packets.

It is necessary to make two remarks concerning the wave function. Firstly the quantities (2.11) and (2.12) involve the wave function at a particular time, whilst the quantity $f(\mathbf{r},\mathbf{p})$ being measured does not depend on time; this formalism is called the Schrödinger picture. We will see later that we could instead have used time independent wave functions or states and time dependent observables; such a formalism is called the Heisenberg picture. Secondly, the wave function must satisfy suitable differentiability conditions in order that $\nabla^2 \psi$ is defined. Thus it is necessary that ψ and $\nabla \psi$ are continuous functions for all $\mathbf{r}$ at each value of time, t. This condition will be important when we consider regions where the potential can change discontinuously.

2.5 The Postulates of Quantum Mechanics *

This section contains a discussion of the states and observables of a single particle in which we try to probe the probability interpretation more deeply. At the same time we will be able to acquire certain mathematical ideas of basic importance in quantum mechanics. As we said in the introduction, the ideas presented here are not all crucial to our later development; the crucial ones are summarized at the end of this section for those students who find this section too difficult.

We have seen that the states of a single particle are represented by wave functions $\psi(\mathbf{r})$ (all at a given time t, which we will not write in explicitly in this section). We require $\int |\psi(\mathbf{r})|^2 d^3\mathbf{r} < \infty$; the set of *all* such wave functions forms a vector space; if ψ and ϕ satisfy this condition so does $(a\psi + b\phi)$, for any complex numbers a and b. The space of all such square integrable functions over three dimensional Euclidean space $\mathbf{R}^3$ is denoted by $L_2(\mathbf{R}^3)$. The dynamical variables for the single particle are represented by operators on this space of wave functions, and have the form $f(\mathbf{r}, (\hbar/i)\nabla)$. We want to restrict ourselves to variables whose average value when measured in any state, ψ, is real, so that

$$\int \psi^* f(\mathbf{r}, (\hbar/i)\nabla)\psi \, d^3\mathbf{r} = \left(\int \psi^* f(\mathbf{r}, (\hbar/i)\nabla)\psi \, d^3\mathbf{r}\right)^*$$
$$= \int \{f(\mathbf{r}, (\hbar/i)\nabla)\psi\}^* \psi \, d^3\mathbf{r} \qquad (2.13)$$

In order for the first and third terms in (2.13) to be equal, it is necessary to change the operation of $f(\mathbf{r}, (\hbar/i)\nabla)$ acting on ψ in the first term to act on the wave function ψ^* in that term. We will consider two separate cases of observables which will cover all the cases we will be interested in. The first case is when the observable f depends only on $\mathbf{r}$ alone. Then the condition (2.13) is equivalent to requiring

$$\int f(\mathbf{r})|\psi(\mathbf{r})|^2 d^3\mathbf{r} = \int f^*(\mathbf{r})|\psi(\mathbf{r})|^2 d^3\mathbf{r} \qquad (2.14)$$

This is evidently satisfied if f is a real function; conversely there are so many states ψ in $L_2(R^3)$, that (2.14) implies that f is real. The other case is when f depends only on the momentum operator $(\hbar/i)\nabla$. Let us first take the very simplest form $f = (\hbar/i)\nabla$. Then (2.11) is

$$\int \psi^*(\hbar/i)\nabla\psi \, d^3\mathbf{r} = \int \{(\hbar/i)\nabla\psi\}^* \psi \, d^3\mathbf{r} \qquad (2.15)$$

We may integrate the left-hand side of (2.15) by parts and use that $|\psi|$ vanishes at $\pm\infty$ (since $\int |\psi(\mathbf{r})|^2 d^3\mathbf{r} < \infty$), so obtain

$$\int \psi^*(\hbar/i)\nabla\psi \, d^3\mathbf{r} = -\int (\hbar/i)\nabla\psi^* \psi \, d^3\mathbf{r}$$

But this is equal to the right hand side of (2.15). Hence (2.13) is satisfied by $f = (\hbar/i)\nabla$. A similar method of integration by parts can be applied to any power of $(\hbar/i)\nabla$, and so to any polynominal in $(\hbar/i)\nabla$ with real coefficients. This is the largest class of functions we need to consider; indeed we will only need to consider the quadratic function $(\hbar/i)^2\nabla^2$, since it has the highest power that arises in the process of quantisation of a single particle by the method of the previous section.

We now consider an important inequality for wave functions. Let ψ, ϕ be two wave functions, and α be a complex number. We consider

$$\int |\phi + \alpha\psi|^2 \, d^3\mathbf{r} = \int \{|\phi|^2 + 2\,\text{Re}(\alpha\phi^*\psi) + |\alpha|^2|\psi|^2\} \, d^3\mathbf{r}$$

(Re denotes the real part of the quantity in parentheses.) We choose α so that $\alpha \int \phi^*\psi \, d^3\mathbf{r}$ is real; to do this we take $\alpha = \lambda e^{-i\mu}$ where $\mu = $ phase of $\int \phi^*\psi \, d^3\mathbf{r}$ and $\lambda = |\alpha|$. Then

$$\int |\phi + \alpha\psi|^2 \, d^3\mathbf{r} = \int |\phi|^2 \, d^3\mathbf{r} + 2\int \lambda\phi^*\psi \, d^3\mathbf{r} + \lambda^2 \int |\psi|^2 \, d^3\mathbf{r} \quad (2.16)$$

But the right hand side of (2.16) must be non-negative, since this is true of the left hand side, so that the quadratic form given by the right hand side of (2.16) must have either complex roots or coinciding real roots. Also, its discriminant must be non-positive, so that

$$\left| \int \phi^*\psi \, d^3\mathbf{r} \right|^2 \leq \left(\int |\phi|^2 \, d^3\mathbf{r} \right) \left(\int |\psi|^2 \, d^3\mathbf{r} \right) \quad (2.17)$$

This is known as Schwarz's inequality and generalizes the inequality for finite sums: $|\sum a_n b_n|^2 \leq (\sum a_n^2)(\sum b_n^2)$. Since in this case the equality can occur in (2.17) if and only if ϕ and ψ are parallel, there must be a complex number a such that $\phi(\mathbf{r}) = a\psi(\mathbf{r})$ for all $\mathbf{r}$ in $\mathbf{R}^3$. The proof of this for Schwarz's inequality for finite sums is straightforward; the proof in our case of functions is interesting in that it shows how close our case is to the finite sum case. To give this proof let us consider the state vector space $L_2(\mathbf{R}^3)$ a little further. In particular we will consider how co-ordinates may be introduced.

In the case of the n-dimensional Euclidean space $\mathbf{R}^n$, we may easily introduce orthogonal co-ordinates by means of the set of n orthogonal normalized basis vectors $\mathbf{e}_1, \cdots, \mathbf{e}_n$ along the n co-ordinate axes. For example in $\mathbf{R}^3$, $\mathbf{e}_1, \mathbf{e}_2$ and $\mathbf{e}_3$ are along the x, y and z co-ordinate axes, with components $\mathbf{e}_1 = (1,0,0)$, $\mathbf{e}_2 = (0,1,0)$, $\mathbf{e}_3 = (0,0,1)$. In $\mathbf{R}^n$, $\mathbf{e}_r$ will have co-ordinates everywhere zero except for one in the rth place. In this case we can express any vector $\mathbf{x}$ as

$$\mathbf{x} = \sum_{r=1}^{n} x_r \mathbf{e}_r$$

$\mathbf{x}$ is then said to be described by its co-ordinates $(x_1, \cdots, x_n)$ in the basis system $\mathbf{e}_1, \cdots, \mathbf{e}_n$. Such a representation of a vector $\mathbf{x}$ is possible in terms of its co-ordinates in any other basis system, say, $\mathbf{f}_1, \cdots, \mathbf{f}_n$,

when these vectors form an orthonormal set, i.e. when $(\mathbf{f}_i, \mathbf{f}_j) = \delta_{ij}$*.
Here the bracket $(\mathbf{f}, \mathbf{g})$ denotes the usual scalar product in $\mathbf{R}^n$ in an abstract form; its concrete form may be obtained in terms of the rectangular co-ordinate system specified by $\mathbf{e}_1, \cdots, \mathbf{e}_n$. If

$$\mathbf{x} = \sum_{r=1}^{n} x_r \mathbf{e}_r = \sum_{r=1}^{n} x'_r \mathbf{f}_r$$

and $|\mathbf{x}|$ is the Euclidean length of the vector from the origin of co-ordinates to $\mathbf{x}$, and if

$$\mathbf{y} = \sum_{r=1}^{n} y_r \mathbf{e}_r = \sum_{r=1}^{n} y'_r \mathbf{f}_r$$

then

$$|\mathbf{x}|^2 = \sum_{r=1}^{n} x_r^2 = \sum_{r=1}^{n} x'^2_r$$

$$(\mathbf{x}, \mathbf{y}) = \sum_{r=1}^{n} x_r y_r = \sum_{r=1}^{n} x'_r y'_r \qquad (2.18)$$

We may now attempt to let n tend to infinity in the above discussion. In this way we obtain an infinite dimensional vector space $\mathbf{S}$ composed of infinite sequences $\mathbf{x} = (x_1, x_2, \cdots)$ with

$$|\mathbf{x}|^2 = \sum_{i \geq 1} x_i^2 < \infty$$

This space $\mathbf{S}$ also has a scalar product between two vectors $\mathbf{x}$ and $\mathbf{y}$ given by

$$(\mathbf{x}, \mathbf{y}) = \sum_{r \geq 1} x_r y_r$$

This is finite, since by Schwarz's inequality (for infinite sums!) $|(\mathbf{x}, \mathbf{y})|^2 \leq |\mathbf{x}|^2 |\mathbf{y}|^2$. If we now let $\mathbf{x}$ have *complex* components, so that

$$|\mathbf{x}|^2 = \sum_{i \geq 1} |x_i|^2, \quad (\mathbf{x}, \mathbf{y}) = \sum_{i \geq 1} x_i^* y_i \qquad (2.19)$$

then the set of all such vectors $\mathbf{x}$ forms a *Hilbert space*, denoted by l_2 (the set of square summable sequences).

The distinguishing features of a Hilbert space is that it is a vector

---

*Where δ_{ij} is the Kronecker delta; $\delta_{ij} = 1$ ($i = j$), $\delta_{ij} = 0$ ($i \neq j$).

space with an *inner product* (ϕ, ψ) between any two elements ϕ, ψ of the space, with the properties:

(i) $(\phi, \phi) \geqslant 0$, $(\phi, \phi) = 0$ if and only if $\phi = 0$ (positivity condition)

(ii) $(a\phi, \psi) = a^*(\phi, \psi)$ (linearity
$(\phi, a\psi + b\chi) = a(\phi, \psi) + b(\phi, \chi)$ conditions)

(iii) $(\phi, \psi) = (\psi, \phi)^*$ (reality condition)

We may evidently regard the space $L_2(\mathbf{R}^3)$ of square integrable wave functions as a Hilbert space with inner product

$$(\phi, \psi) = \int \phi^*(\mathbf{r}) \psi(\mathbf{r}) d^3\mathbf{r} \qquad (2.20)$$

This inner product evidently satisfies the conditions (i), (ii), (iii) stated above; we may also introduce an *orthonormal* set of functions $\{\phi_n(\mathbf{r})\}$ with

$$\int \phi_n^*(\mathbf{r}) \phi_m(\mathbf{r}) d^3\mathbf{r} = (\phi_n, \phi_m) = \delta_{nm} \qquad (2.21)$$

so that any ϕ in $L_2(\mathbf{R}^3)$ can be expressed as

$$\phi(\mathbf{r}) = \sum a_n \phi_n(\mathbf{r}) \qquad (2.22)$$

where

$$a_n = \int \phi(\mathbf{r}) \phi_n^*(\mathbf{r}) d^3\mathbf{r} \qquad (2.23)$$

The quantities a_n may be regarded as the infinite set of co-ordinates of the wave function ϕ, and the length of ϕ, which we denote by $\|\phi\|$ and define as

$$\|\phi\| = (\phi, \phi)^{\frac{1}{2}} = \left\{ \int |\phi(\mathbf{r})|^2 d^3\mathbf{r} \right\}^{\frac{1}{2}} \qquad (2.24)$$

satisfies $\|\phi\|^2 = \sum |a_n|^2$, as we can see by evaluating the right hand side of (2.24) by means of (2.22), using (2.21) and (2.23). Thus we can consider the space of wave functions as the space l_2 of square-summable sequences.

Let us return to the Schwarz inequality (2.17). We choose an orthonormal set of wave functions ϕ_n, with ϕ proportional to ϕ_1, so that $\phi = \alpha \phi_1$, $\alpha = \|\phi\|$. This choice may always be made; indeed we may start from $\phi_1 = \phi / \|\phi\|$ and always choose the remaining $\phi_n (n \neq 1)$ suitably orthogonal to each other and ϕ_1, and then normalize them. Suppose that ψ can be expressed as

$$\psi = \sum_{n \geq 1} a_n \phi_n$$

so that
$$(\phi, \psi) = \sum a_n \alpha(\phi_1, \phi_n) = a_1 \alpha.$$
Then
$$|(\phi, \psi)| = |a_1| \, \|\phi\|$$

In order that the Schwarz inequality be an equality we need

$$\|\phi\| |a_1| = \|\phi\| \, \|\psi\| = \|\phi\| \{ \sum_{n \geq 1} |a_n|^2 \}^{\frac{1}{2}} \tag{2.25}$$

This can only be achieved if $a_n = 0$ $(n > 1)$, so that $\psi = a_1 \phi_1 \propto \phi$, as we wished to show.

Before we leave this discussion of the space of wave functions in terms of Hilbert space theory, it is useful to give a brief discussion of operators on a Hilbert space; this will allow us to discuss certain properties of many different operators at once. Let us consider the concrete Hilbert space $L_2(\mathbf{R}^3)$ (though any other Hilbert space will give similar results). A linear operator T is defined on $L_2(\mathbf{R}^3)$ as a linear mapping which takes each ϕ in $L_2(\mathbf{R}^3)$ into $T(\phi)$ in $L_2(\mathbf{R}^3)$ so that

$$T(a\phi + b\psi) = a\, T(\phi) + b\, T(\psi)$$

for any complex numbers a, b. An important linear operator is $(\hbar/i)\nabla$, as is $(-\hbar^2 \nabla^2/2m) + V(\mathbf{r})$, the Hamiltonian operator. We define ϕ_λ to be an *eigenvector* of T belonging to the eigenvalue λ if

$$T(\phi_\lambda) = \lambda \phi_\lambda \tag{2.26}$$

In other words, when T is applied to ϕ_λ a multiple of ϕ_λ results. The concept of eigenvalue and eigenvector is extremely important in quantum mechanics, and familiarity with it is very necessary. For example $e^{i\mathbf{k} \cdot \mathbf{r}}$ is an eigenvector of the momentum operator $(\hbar/i)\nabla$, belonging to the eigenvalue $\hbar\mathbf{k}$:

$$(\hbar/i)\nabla(e^{i\mathbf{k} \cdot \mathbf{r}}) = \hbar\mathbf{k} e^{i\mathbf{k} \cdot \mathbf{r}}$$

(Such a plane-wave eigenvector is not square-summable, so is not an eigenvector in the strict sense we defined above. However, we can use such an eigenvector provided we are careful about it; whenever we do use such plane waves we will also use care with them.)

Another operator we have already discussed is multiplication by $\mathbf{r}$, say by x, the first component of $\mathbf{r}$. The eigenvector of x is a very important 'function', the δ-function, $\delta(x-a)$ say, which has the property that $\delta(x) = 0$ if $x \neq 0$ and $\delta(x)$ is infinite at $x = 0$ in such a way that for any function $\phi(x)$ which is continuous near $x = 0$ then $\int \delta(x)\phi(x)\,dx = \phi(0)$. Of course $\delta(x)$ is not in $L_2(\mathbf{R}^3)$, but may be regarded as the limit of a suitable sequence of square integrable functions $\phi_n(x)$, say

$$\phi_n(x) = \tfrac{1}{2}ne^{-n|x|}$$

Evidently

$$x\,\delta(x-a) = a\,\delta(x-a) \tag{2.27}$$

since the left hand side of (2.27) is only non-zero for $x = a$.

The δ-function was originally introduced by Dirac and was made respectable by Laurent Schwarz 20 years later. We will not consider how to make it respectable, but use it intuitively. We can relate $\delta(x-a)$ and e^{ikx} very naturally and also relate the operators x and $(\hbar/i)\,d/dx$ by the *Fourier transform*. This is defined for any wave function ϕ in $L_2(R)$ as the function

$$\tilde{\phi}(k) = F\phi(k) = \frac{1}{\sqrt{(2\pi)}} \int_{-\infty}^{+\infty} e^{ikx}\phi(x)\,dx \tag{2.28}$$

We may invert the Fourier transformation (2.28) to obtain

$$\phi(x) = \frac{1}{\sqrt{(2\pi)}} \int_{-\infty}^{+\infty} e^{-ikx}\tilde{\phi}(k)\,dk = F^{-1}\tilde{\phi}(k) \tag{2.29}$$

To prove the validity of (2.29), we need to prove that when we replace $\tilde{\phi}(k)$ by the right hand side of (2.28), the right hand side of (2.29) correctly gives $\phi(x)$, or that

$$\phi(x) = \frac{1}{2\pi} \int_{-\infty}^{+\infty} dk\, e^{-ikx} \int_{-\infty}^{+\infty} dy\, e^{iky}\phi(y)$$

or that

$$\frac{1}{2\pi} \int_{-\infty}^{+\infty} e^{ik(y-x)}\,dk = \delta(y-x) \tag{2.30}$$

To prove this equation we modify the integrand so that it is damped for large k, replacing it by $e^{ik(y-x)-\alpha^2 k^2}$, with α positive. Then to prove (2.30) we need to show that when each side is multiplied by an arbitrary (smooth) function $g(y)$, integrated over all y, and then α made to approach zero we obtain $g(x)$, or that

$$\lim_{\alpha \to 0} \frac{1}{2\pi} \int_{-\infty}^{+\infty} e^{ik(y-x) - \alpha^2 k^2} g(y) \, dy \, dk = g(x) \qquad (2.31)$$

But on the left hand side of (2.31) the k-integral gives, by a suitable change of variable $\alpha k - (i/2\alpha)y = z$,

$$\int_{-\infty}^{+\infty} e^{iky - \alpha^2 k^2} \, dk = \frac{1}{\alpha} e^{-y^2/4\alpha^2} \int_{-\infty}^{+\infty} e^{-z^2} \, dz$$

$$= \alpha^{-1} e^{-y^2/4\alpha^2} \sqrt{\pi}$$

Now the functions $\alpha^{-1} e^{-y^2/4\alpha^2}$ tend to zero as α tends to zero if y is non-zero, whilst they become infinite if $y \neq 0$, in such a way that

$$\int_{-\infty}^{+\infty} \alpha^{-1} e^{-y^2/4\alpha^2} \, dy = 2\sqrt{\pi}$$

This shows that

$$\lim_{\alpha \to 0} \frac{1}{2\pi} \int_{-\infty}^{+\infty} e^{ik(y-x) - \alpha^2 k^2} \, dk = \delta(y-x) \qquad (2.32)$$

so proving (2.31) and (2.30), and also the Fourier inversion formula (2.29). We regard (2.30) as a shorthand (though not completely defined) version of (2.31) (which is well defined). Thus the Fourier transform of a plane wave is a δ-function in momentum space; this is to be expected, since in momentum space the δ-function is the eigenfunction of the momentum operator, which is now multiplication by $\hbar k$, because

$$F((\hbar/i) \, d\phi/dx) = \hbar k \tilde{\phi}(k)$$

as can be seen directly by differentiating the right hand side of (2.28).

A very important property of the Fourier transformation is that it preserves the *length* of the wave function:

$$\int_{-\infty}^{+\infty} |\phi(x)|^2 \, dx = \int_{-\infty}^{+\infty} |\tilde{\phi}(k)|^2 \, dk \qquad (2.33)$$

Thus we may interpret $|\tilde{\phi}(k)|^2 \, dk$ as the relative probability of finding the particle momentum to be in the interval $(\hbar k, \hbar(k+dk))$. The momentum space analogue of the plane wave (for which $|\phi(x)| \equiv 1$) is the function ψ with $|\tilde{\psi}(k)| \equiv 1$; if we take $\tilde{\psi}(k) = e^{ikx_0}$ then we see $\psi(x) \propto \delta(x - x_0)$. Thus the function $\delta(x - x_0)$ is the eigenvector of the position operator x with eigenvalue x_0; in momentum space it describes a state e^{ikx_0} in which the particle has momentum distributed with uniform probability for all possible values. This means that there is infinite uncertainty in the momentum when there is zero uncertainty in the position of the particle. The converse case of zero momentum uncertainty and infinite position uncertainty arises for the wave function $e^{ik_0 x}$ or momentum space wave function $\delta(k-k_0)$. These are the two extremes of the *uncertainty principle* of Heisenberg: the more uncertainty in position, the less uncertainty in momentum for a given state of a particle, and conversely. We will discuss this in more detail in the next section. In the meantime we will generalize the probability interpretation a little further.

Let us return to the operator T and its eigenvalues and eigenvectors of (2.26). We can suppose that the set of eigenvalues λ are discrete, so can be labelled $\lambda_1, \lambda_2, \cdots$, and the corresponding states denoted by $\phi_1, \phi_2, \cdots$. We further suppose that T is *self adjoint*, or that:

$$(T\phi, \psi) = (\phi, T\psi) \qquad (2.34)$$

for any state vectors ϕ, ψ. We have already seen that $(\hbar/i)\nabla$ and $\{(-\hbar^2 \nabla^2/2m) + V(\mathbf{r})\}$ are self-adjoint operators. A self-adjoint operator T has two very important properties:

(a) Any eigenvalue λ of T is real, since if $T\phi_\lambda = \lambda \phi_\lambda$ then

$$(T\phi_\lambda, \phi_\lambda) = (\lambda \phi_\lambda, \phi_\lambda) = \lambda^* \|\phi_\lambda\|^2$$
$$(T\phi_\lambda, \phi_\lambda)^* = (\phi_\lambda, T\phi_\lambda) = \lambda \|\phi_\lambda\|^2$$

But $(T\phi_\lambda, \phi_\lambda) = (\phi_\lambda, T\phi_\lambda)$ if T is self adjoint, so $\lambda = \lambda^*$ and λ is real.

(b) If $\lambda \neq \lambda'$, then $(\phi_\lambda, \phi_{\lambda'}) = 0$, for, since T is self adjoint,

$$(T\phi_\lambda, \phi_{\lambda'}) = (\phi_\lambda, T\phi_{\lambda'}) \qquad (2.35)$$

and the left hand side of (2.35) is $\lambda(\phi_\lambda, \phi_{\lambda'})$ (which is real), whilst the right hand side is $\lambda'(\phi_\lambda, \phi_{\lambda'})$, so that if $\lambda \neq \lambda'$ we must have $(\phi_\lambda, \phi_{\lambda'}) = 0$.

We suppose that the set $\{\phi_n\}$ of eigenvectors of T is a *complete* orthonormal set of vectors, so that for any state vector ϕ we have the expansion

$$\phi = \sum_n a_n \phi_n, \quad \|\phi\|^2 = \sum |a_n|^2, \qquad a_n = (\phi, \phi_n) \qquad (2.36)$$

Let us choose $\|\phi\| = 1$. Then the average value of the dynamical variable in the state ϕ is

$$\langle T \rangle = (\phi, T\phi) = \left(\sum_n a_n \phi_n, T \sum_m a_m \phi_m \right)$$

$$= \left(\sum_n a_n \phi_n, \sum_m \lambda_m a_m \phi_m \right)$$

$$= \sum_{n,m} a_n^* a_m \lambda_m (\phi_n, \phi_m) = \sum_n |a_n|^2 \lambda_n \qquad (2.37)$$

Thus the average value of T in the state ϕ is a weighted sum of the values of T in the states ϕ_n; the weight for the nth term is $|a_n|^2$. We may interpret $|a_n|^2$ as the probability that the measurement of T in the state ϕ gives the value λ_n, which is certainly consistent with (2.37). We thus have an extension of the probability interpretation to any self-adjoint operator T on the state-vector space, where T is supposed to represent a suitable dynamical variable for the particle; the possible values of T are its eigenvalues, and their probability of occurring are given by the squared *projections* $|(\phi, \phi_n)|^2$ of ϕ on the state ϕ_n.

This extension of the probability interpretation to dynamical variables described by self-adjoint operators is extremely important in the further development of quantum mechanics; in this book we will restrict ourselves to the simpler cases of the position and momentum operators $\mathbf{r}$, $(\hbar/i)\nabla$, and to Hamiltonian operators of the form $H = (-\hbar^2 \nabla^2/2m) + V(\mathbf{r})$.

Summary of § 2.5

(i) The wave functions describing the possible states of a single particle are functions $\psi(\mathbf{r})$ of the position vector $\mathbf{r}$ of the particle, with $\int |\psi(\mathbf{r})|^2 \, d^3\mathbf{r} < \infty$, the integral being taken over all space. These functions are *square integrable*.

(ii) The dynamical variables are the functions of $\mathbf{r}$ and $(\hbar/i)\nabla$ obtained by replacing $\mathbf{p}$ by $(\hbar/i)\nabla$ in the classical dynamical variable $f(\mathbf{r}, \mathbf{p})$. The average value of $f(\mathbf{r}, \mathbf{p})$ in the state $\psi(\mathbf{r})$ is defined to be

$$\langle f \rangle = \int \psi^*(\mathbf{r}) f(\mathbf{r}, (\hbar/i)\nabla) \psi(\mathbf{r}) \, d^3\mathbf{r} \bigg/ \int |\psi(\mathbf{r})|^2 \, d^3\mathbf{r}$$

(iii) We may define an inner product between two such states to be the complex number (ϕ, ψ) given by

$$(\phi,\psi) = \int \phi^*(\mathbf{r})\psi(\mathbf{r}) \, d^3\mathbf{r}$$

This satisfies the Schwarz inequality:

$$|(\phi,\psi)|^2 \leq \left(\int |\phi(\mathbf{r})|^2 \, d^3\mathbf{r} \right) \left(\int |\psi(\mathbf{r})|^2 \, d^3\mathbf{r} \right)$$

with equality if and only if ϕ is proportional to ψ. The average value of the dynamical variable in the state ψ is then $(\psi, f\psi)$, and is required to be real. This will be true if, for all state vectors ψ,

$$(\psi, f\psi) = (f\psi, \psi)$$

This condition will automatically be satisfied by all observables we discuss in this book; examples are $(\hbar/i)\nabla$ and $\{(-\hbar^2\nabla^2/2m) + V(\mathbf{r})\}$.

(iv) A *linear operator* T acts on a wave function ψ in a linear fashion to produce another wave function. For example, $(\hbar/i)\nabla$ and $\{(-\hbar^2\nabla^2/2m) + V(\mathbf{r})\}$ are linear operators. The wave function ϕ_λ is an *eigenvector* of T belonging to the eigenvalue λ if

$$T\phi_\lambda = \lambda\phi_\lambda$$

The probability that the measurement of T in the state ψ gives the value λ is $|(\psi,\phi_\lambda)|^2$. If the particle has the wave function ϕ_λ, then T certainly has the value λ.

(v) The Fourier transform $\tilde{\phi}(k)$ of $\phi(x)$ is defined as

$$\tilde{\phi}(k) = \frac{1}{\sqrt{(2\pi)}} \int_{-\infty}^{+\infty} e^{ikx} \phi(x) \, dx$$

which may be inverted to give

$$\phi(x) = \frac{1}{\sqrt{(2\pi)}} \int_{-\infty}^{+\infty} e^{-ikx} \tilde{\phi}(k) \, dk$$

Then the momentum operator $(\hbar/i)d/dx$ becomes $\hbar k$ under Fourier transformation, so k-space is called momentum space; $|\tilde{\phi}(k)|^2 \, dk$ is the relative probability that the momentum of the particle is in the interval $(\hbar k, \hbar(k+dk))$, if the particle has wave function $\phi(x)$.

(vi) The δ-function is that function $\delta(x)$ which vanishes unless x is zero, and then is infinite in such a way at $x = 0$ that the area under the graph of $\delta(x)$ is unity:

$$\delta(x) = 0 \quad (x \neq 0)$$
$$\int_{-\infty}^{+\infty} \delta(x)\, dx = 1$$

For any smooth function $g(x)$,
$$\int_{-\infty}^{+\infty} \delta(x) g(x)\, dx = g(0)$$

Hence $\delta(x)$ is the Fourier transform of a plane wave
$$\delta(x) = \frac{1}{2\pi} \int_{-\infty}^{+\infty} e^{ikx}\, dk$$

with $\delta(x-a)$ an eigenvector of the position operator with eigenvalue a:
$$x\,\delta(x-a) = a\,\delta(x-a)$$

Hence a particle described by a δ function $\delta(x-a)$ is certainly at the point a; since its Fourier transform is e^{ika} then it has equal probability of its momentum being in any interval of length $\hbar dk$. Conversely the wave function $e^{ik_0 x}$ describes a particle equally spread throughout space, but with Fourier transform $\delta(k-k_0)$, so describing a particle with certain momentum $\hbar k_0$.

The statements in (v) and (vi), appropriate to one-dimensional situations have a straightforward generalisation to three dimensions.

2.6 The Uncertainty Principle

We discussed in the previous section two extreme wave functions for a single particle, the δ-function wave function $\delta(x-x_0)$ in which the particle is *always* at x_0 with indefinite momentum, and the plane wave function $e^{ik_0 x}$ in which the particle has indefinite position but momentum certain to be $\hbar k_0$. We now show that in any wave function the uncertainty Δx in position and the uncertainty Δp in momentum can never both be zero, and indeed must satisfy the Heisenberg uncertainty relation
$$\Delta x \, \Delta p \geqslant \hbar/2 \tag{2.38}$$

Let us show this for any wave function ψ (where we will take one-dimensional motion, for simplicity). We define Δx and Δp as the root mean square uncertainties in x and p:
$$(\Delta x)^2 = \int |\psi(x)|^2 (x - \langle x \rangle)^2 \, dx \tag{2.39}$$

and
$$(\Delta p)^2 = \int |(p-\langle p\rangle)\psi|^2 \, dx \tag{2.40}$$
where
$$\langle x \rangle = \int x|\psi|^2 \, dx \tag{2.41}$$
$$\langle p \rangle = \int (p\psi)^*\psi \, dx \tag{2.42}$$

and as usual $p = (\hbar/i)(d/dx)$, $\int |\psi|^2 \, dx = 1$. We may always choose our origin of co-ordinates so that $\langle x \rangle = 0$; thus if the new co-ordinate is a distance a from the old we take $a = \int x|\psi|^2 dx / \int |\psi|^2 \, dx$ so that $\langle x' \rangle = 0$ if $x' = (x-a)$ is the new co-ordinate. We may similarly choose $\langle p \rangle = 0$ by a suitable choice of origin in momentum space, which is equivalent to choosing the phase of ψ suitably: thus if we replace $\psi(x)$ by $\psi'(x) = e^{-i\langle p\rangle x/\hbar}\psi(x)$ then the value of $\langle x \rangle$ and Δx remain unchanged whilst the new value $\langle p \rangle'$ of $\langle p \rangle$ is zero, since

$$\langle p \rangle = \langle p \rangle \int |\psi'|^2 \, dx + \langle p \rangle' = \langle p \rangle + \langle p \rangle'.$$

Thus we may assume, without loss of generality, that $\langle x \rangle = \langle p \rangle = 0$ for ψ. To derive (2.38) let us consider the quantity

$$-\int (\hbar/i)(d\psi^*/dx)x\psi \, dx$$

Integrating this by parts we obtain

$$-\int \frac{\hbar}{i} \frac{d\psi^*}{dx} x\psi \, dx = \frac{\hbar}{i} \int |\psi|^2 \, dx + \frac{\hbar}{i} \int \frac{d\psi}{dx} x\psi^* \, dx$$

where the contributions from the end points of the integration vanish. We may rearrange this to be

$$-\int \frac{\hbar}{i} \frac{d\psi^*}{dx} x\psi \, dx - \left(\int -\frac{\hbar}{i} \frac{d\psi^*}{dx} x\psi \, dx\right)^* = \frac{\hbar}{i} \int |\psi|^2 \, dx$$

or

$$2i \, \text{Im}\left\{-\frac{\hbar}{i} \int \frac{d\psi^*}{dx} x\psi \, dx\right\} = \frac{\hbar}{i} \int |\psi|^2 \, dx$$

where Im denotes the imaginary part of the quantity in the brackets.

Then, squaring moduli of both sides of this equation, we obtain

$$\hbar^2 \left\{ \int |\psi|^2 \, dx \right\}^2 = 4 \left| \text{Im} \left\{ \int -(\hbar/i)(d\psi^*/dx) x \psi \, dx \right\} \right|^2 \quad (2.43)$$

We may increase the right hand side of (2.43) by removing the Im, since $|\text{Im}\, z| \leq |z|$ for any complex number z. By the Schwarz inequality

$$\left| \int -\frac{\hbar}{i} \frac{d\psi^*}{dx} x\psi \, dx \right|^2 \leq \left(\int \left| \frac{\hbar}{i} \frac{d\psi^*}{dx} \right|^2 dx \right) \left(\int |x\psi|^2 \, dx \right) \quad (2.44)$$
$$= (\Delta x)^2 (\Delta p)^2$$

Combining this with (2.43) we have

$$\hbar^2/4 \leq (\Delta x)^2 (\Delta p)^2$$

or

$$\Delta x \, \Delta p \geq \hbar/2.$$

which is the Heisenberg uncertainty principle, (2.38).

The uncertainty principle is extremely important in giving us a limitation on the accuracy with which the *simultaneous* measurement of the position x and momentum p of a particle can be measured. It states that the higher the accuracy in the measurement of one of these quantities, the lower will be the accuracy in the measurement of the other quantity. This is very different from the classical particle for which both x and p could be measured with arbitrary accuracy. We may trace the origin of this inaccuracy in quantum mechanics to the fact that x and p are to be interpreted as operators which *do not commute*, that is the products xp and px differ:

$$px - xp = \left(\frac{\hbar}{i} \frac{d}{dx} \right) x - x \left(\frac{\hbar}{i} \frac{d}{dx} \right) = -i\hbar \quad (2.45)$$

(where both sides of (2.45) are to be interpreted as operators acting on wave functions). In classical mechanics $px = xp$; we may interpret (2.45) in terms of the difference obtained between a measurement of p and then immediately afterwards of x, and a measurement of x and then p. This difference, which vanishes in classical mechanics, gives rise in quantum mechanics to simultaneous uncertainty in x and p, as expressed by the uncertainty principle.

We have already considered two limiting cases of the uncertainty relation (2.38):

(a) a plane wave, $\psi(x) = e^{ik_0 x}$, with momentum space wave function $\tilde{\psi}(k) = \delta(k_0 - k)$, for which $\Delta x = \infty$, $\Delta p = 0$.
(b) a δ-function, $\psi(x) = \delta(x - x_0)$, with momentum space wave function $\tilde{\psi}(k) = e^{ikx_0}$, for which $\Delta x = 0$, $\Delta p = \infty$.

The third limiting case of interest is that for which there is minimum uncertainty jointly in Δx and Δp so that the equality holds in (2.38): $\Delta x \Delta p = \frac{1}{2}\hbar$. Such a possibility only arises for a unique wave function; to see this we turn back to the proof of the Schwarz inequality which we gave in the previous section.

We see that the inequality in (2.44) becomes an *equality* provided that $(\hbar/i)(d\psi/dx)$ and $x\psi$ are parallel, so that $d\psi/dx = Cx\psi$, where C is some constant. We still end up with an inequality in (2.38) unless the replacement of the imaginary part on the right-hand side of (2.43) by the whole complex number $\int -(\hbar/i)(d\psi^*/dx)x\psi \, dx$ in the modulus sign causes an increase. This increase won't happen if z is purely imaginary; when $d\psi/dx = Cx\psi$ then $z = i\hbar C \int x^2 |\psi|^2 \, dx$, so we need C real. But then $\psi = Ne^{-Cx^2/2}$, which is a Gaussian wave function. This decreases very rapidly for larger x. N is a normalizing constant, chosen so that $\int_{-\infty}^{+\infty} |\psi|^2 \, dx = 1$, and since $\int_{-\infty}^{+\infty} e^{-x^2} \, dx = \sqrt{\pi}$ then $N = (C/\pi)^{\frac{1}{4}}$. We see that a Gaussian wave function has a momentum space wave function which is also a Gaussian:

$$\tilde{\psi}(k) = \frac{1}{\sqrt{(2\pi)}} N \int_{-\infty}^{+\infty} e^{ikx - Cx^2/2} \, dx$$

$$= \frac{1}{\sqrt{(2\pi)}} N e^{-k^2/2C} \int_{-\infty}^{+\infty} e^{-C/2(x - ik/C)^2} \, dx$$

$$= (\pi C)^{-\frac{1}{4}} e^{-k^2/2C} \qquad (2.46)$$

This Gaussian has $1/C$ replaced by C; this is to be expected since the 'spread' of position about $x = 0$ in the wave function e^{-Cx^2} is given by

$$(\Delta x)^2 = N^2 \int_{-\infty}^{+\infty} x^2 e^{-Cx^2} \, dx = (1/2C)$$

Hence $\Delta x = 1/\sqrt{(2C)}$; similarly $\Delta p = \hbar \sqrt{C/2}$, so $\Delta x \Delta p = \hbar/2$, as we expect.

For the sake of completeness we note that the Gaussian wave function centred at $\langle x \rangle$ and with mean momentum $\langle p \rangle$ is

$$Ne^{(i/\hbar)\langle p \rangle x - C(x-\langle x \rangle)^2/2}$$

In three dimensions this wave function becomes $Ne^{(i/\hbar)\langle \mathbf{p} \rangle \cdot \mathbf{x} - C(\mathbf{x}-\langle \mathbf{x} \rangle)^2/2}$, which is just a product of Gaussian wave functions in each coordinate separately.

It is possible to give a physical interpretation to the uncertainty principle by means of the Gamma ray microscope of Heisenberg; this can only be constructed in thought, but shows the physical nature of the limitations of the uncertainty principle. In the microscope a beam of light is shone onto an electron at right angles to its motion, as in Fig. 2.7. In order to localize the electron as well as possible along its

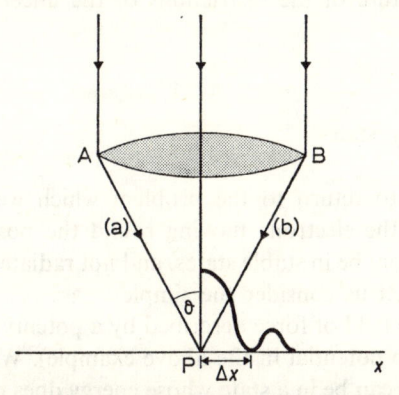

Fig. 2.7 A diagrammatic representation of Heisenberg's X-ray microscope. The lens at AB focuses light onto the particle at P; the rays (a), (b) are the extreme rays giving the largest errors in the measurement of the x-component p_x of the momentum of the particle; the curve shows the first diffraction maximum away from the central position.

line of motion we try to obtain as narrow a beam of light as possible. However, the resolving power of any optical instrument is limited, so that the accuracy of measurement of the position of the electron is also limited. This limit may be obtained by considering the interference pattern produced by a lens as the angle θ subtended by the lens at the object is made smaller. The accuracy Δx of measurement of the position of the particle is always greater than that between the peak and the first minimum of this pattern. By the arguments used in §2.3, $\Delta x \geqslant \lambda/2\sin\theta$.

The accuracy of the measurement of the x-momentum of the particle is limited by the momentum given to the particle by a single photon which is scattered from it. The maximum value of such momentum occurs for photons travelling along the directions a or b in Fig. 2.7; such photons carry momentum h/λ, so their maximum x-component of momentum is $p_x = h\sin\theta/\lambda$; this is equal to the error Δp_x in the x-component of momentum for the particle. Thus

$$\Delta x \Delta p_x \geqslant h/2$$

which is roughly the Heisenberg uncertainty limit. This 'thought experiment' cannot be regarded as a justification of the uncertainty principle but should be regarded as an example of the way in which the physical nature of the restrictions of the uncertainty principle arise.

2.7 Stationary States

We want now to return to the problem which we considered in §2.1: how can the electrons moving round the positively charged nucleus in an atom be in stable states, and not radiate all their energy very rapidly? Let us consider the simplest case of a single particle moving in some field of force described by a potential (which would be the Coulomb potential in the above example). We wish to show that the particle can be in a state whose energy does not change with time. If we can find such a state, then we can use it to describe the stable state of an electron in an atom, and so solve the problem of the stability of matter raised in §2.1.

We take the classical Hamiltonian for the particle to be

$$H = (\mathbf{p}^2/2m) + V(\mathbf{r});$$

by our discussion of §2.4 we may set up the Schrödinger equation (2.6') to describe the time development of the wave function for the particle in quantum mechanics:

$$i\hbar \frac{\partial \psi}{\partial t} = \left\{ -\frac{\hbar^2 \nabla^2}{2m} + V(\mathbf{r}) \right\} \psi \qquad (2.6')$$

where ψ is a function of the position of the particle **r** and the time t. Let us take a time independent wave function $\psi(\mathbf{r})$ which is an eigenvector of H belonging to the eigenvalue E:

$$H\psi(\mathbf{r}) = \left\{-\frac{\hbar^2 \nabla^2}{2m} + V(\mathbf{r})\right\}\psi(\mathbf{r}) = E\psi(\mathbf{r}) \quad (2.47)$$

We can use this wave function to write down a solution of the time dependent Schrödinger equation (2.6'):

$$\psi(\mathbf{r}, t) = e^{-iEt/\hbar}\psi(\mathbf{r}) \quad (2.48)$$

since the wave function (2.48) satisfies

$$i\hbar\frac{\partial \psi}{\partial t} = E\psi(\mathbf{r}, t) = H\psi(\mathbf{r}, t) \quad (2.49)$$

the last equality in (2.49) coming from (2.47). Thus we have obtained the time dependent wave function (2.48); evidently at any time t, the state $\psi(\mathbf{r}, t)$ is an eigenstate of H, the exponential function not producing any effect when H is applied to $e^{-iEt/\hbar}\psi(\mathbf{r})$. This means that the value of the energy is *always* E when measured in the state described by the wave function $\psi(\mathbf{r}, t)$ of (2.48); this state is thus the stationary state we are looking for.

We note that it is usual to require that the solution $\psi(\mathbf{r})$ of the eigenvalue problem (2.47) is localized in the sense that $|\psi(\mathbf{r})|^2$ vanishes for large $|\mathbf{r}|$ fast enough for $\int |\psi(\mathbf{r})|^2 d^3\mathbf{r} < \infty$. This would not be the case for the plane wave $e^{i\mathbf{k}\cdot\mathbf{r}}$ with $V = 0$, so that $E = \hbar^2 k^2/2m$. The plane wave is evidently not localized and is not expected to correspond to the stable state describing an electron localized near the nucleus.

The stable states for which $\int |\psi(\mathbf{r})|^2 d^3\mathbf{r} < \infty$, in other words the normalizable states, may be regarded as *bound states* in the potential V. It is usual that the possible values of E for the various possible normalizable solutions of (2.47) form a discrete set of values. This is certainly not the case for the values of E corresponding to the non-normalizable states; for them the range of possible values of E is usually continuous (being all the positive real axis for the free particle case with $E = \hbar^2 k^2/2m$). These non-normalizable states are usually called *scattering states*; they correspond to particles which are moving in plane wave states for large values of $|\mathbf{r}|$, where $V(\mathbf{r}) \sim 0$. We will return to these scattering states when we discuss scattering theory in Chapter 6, and also when we discuss one dimensional problems in the next chapter; other than in those places we will only consider states describable by normalizable wave functions.

We now see that the main problem of quantum mechanics appears to be solving the time independent Schrödinger equation (2.47) for the energies E and the related eigenfunctions. This will take up most of our time in the remainder of this book. Before we turn to this we will briefly discuss the important concept of *probability current*. This may be defined by means of the time dependent Schrödinger equation (2.6') and its complex conjugate:

$$(-\hbar^2/2m)\nabla^2\psi + V\psi = i\hbar\, \partial\psi/\partial t \qquad (2.50)$$

$$(-\hbar^2/2m)\nabla^2\psi^* + V\psi^* = -i\hbar\, \partial\psi^*/\partial t \qquad (2.51)$$

where we are using that V is real.

If we multiply (2.50) by ψ^* and substract (2.51) multiplied by ψ we obtain

$$-\frac{\hbar^2}{2m}\{\psi^*\nabla^2\psi - (\nabla^2\psi^*)\psi\} = i\hbar\left\{\psi^*\frac{\partial\psi}{\partial t} + \psi\frac{\partial\psi^*}{dt}\right\} = i\hbar\frac{\partial}{\partial t}(|\psi|^2)$$

Rewriting the left hand side of this equation slightly we obtain

$$\partial/\partial t\,(|\psi|^2) + (\hbar/2mi)\nabla.(\psi^*\nabla\psi - \psi\nabla\psi^*) = 0 \qquad (2.52)$$

or

$$\partial\rho/\partial t + \nabla.\mathbf{S} = 0 \qquad (2.53)$$

where

$$(\rho, \mathbf{S}) = (|\psi|^2, (\hbar/2mi)(\psi^*\nabla\psi - \psi\nabla\psi^*)) \qquad (2.54)$$

We may interpret (2.53) as an equation of continuity, and use it to obtain

$$\frac{\partial}{\partial t}\int_V \rho\, dV = -\int_V \nabla.\mathbf{S}\, dV = -\int_S \mathbf{S}.d\boldsymbol{\sigma} \qquad (2.55)$$

where S is the surface bounding the volume V, and $d\boldsymbol{\sigma}$ is the vector element of area on S pointing outwards. We may interpret (2.55) as follows: the left hand side is the rate of increase of probability of finding the particle inside the volume V, so that if we assume probability is conserved, then this rate of increase is due to the inflow of

THE DEVELOPMENT OF WAVE MECHANICS

probability across S into V. This inflow is given by **S**, which is thus called the *probability current*.

This current has very natural properties. In particular, for a plane wave

$$\psi(\mathbf{r}) = e^{i\mathbf{p}\cdot\mathbf{r}/\hbar}$$

We have $\mathbf{S} = \mathbf{p}/m = \mathbf{v}$ which is the velocity of a 'particle' moving with mass m and momentum **p**. This is exactly the rate of flow of probability expected for a plane wave. Also, for a stationary state $\rho(\mathbf{r}, t) = |\psi|(\mathbf{r})|^2$ and so is independent of time (as it should be for a *stationary* state).

2.8 The Motion of Wave Packets *

To conclude this chapter we will see that the average position and momenta of a particle in quantum mechanics satisfy the classical equations of motion:

$$\frac{d}{dt}\langle \mathbf{r} \rangle = \langle \mathbf{p} \rangle/m, \qquad \frac{d}{dt}\langle \mathbf{p} \rangle = -\langle \nabla V \rangle \qquad (2.56)$$

Suppose $\psi(\mathbf{r}, t)$ satisfies the time dependent Schrödinger equation

$$-\left(\frac{\hbar^2}{2m}\right)\nabla^2\psi + V\psi = i\hbar\left(\frac{\partial \psi}{\partial t}\right) \qquad (2.57)$$

and consider

$$\frac{d}{dt}\langle \mathbf{r} \rangle = \frac{d}{dt}\int \psi^*(\mathbf{r}, t)\mathbf{r}\,\psi(\mathbf{r}, t)\,d^3\mathbf{r}$$

$$= \int \left\{ \mathbf{r}\psi\left(\frac{\partial \psi^*}{\partial t}\right) + \psi^*\mathbf{r}\left(\frac{\partial \psi}{\partial t}\right)\right\} d^3\mathbf{r}$$

We use (2.57) to obtain

$$\frac{d}{dt}\langle \mathbf{r}\rangle = \left(\frac{1}{i\hbar}\right)\int d^3\mathbf{r}\left\{\left(\frac{\hbar^2}{2m}\nabla^2\psi^* - V\psi^*\right)\mathbf{r}\psi + \psi^*\mathbf{r}\left(V\psi - \frac{\hbar^2}{2m}\nabla^2\psi\right)\right\}$$

$$= \left(\frac{1}{i\hbar}\right)\left(\frac{-\hbar^2}{2m}\right)\int \mathbf{r}(\psi^*\nabla^2\psi - (\nabla^2\psi^*)\psi)\,d^3\mathbf{r}$$

$$= \frac{\hbar}{m} \text{Im} \left\{ \int \mathbf{r}\psi^*(\nabla^2\psi) \, d^3\mathbf{r} \right\} \qquad (2.58)$$

Also

$$\sum_{i=1}^{3} \partial/\partial r_i (\mathbf{r}\psi^* \partial/\partial r_i \psi) = \mathbf{r}(\nabla\psi^* \cdot \nabla\psi) + \mathbf{r}\psi^* \nabla^2\psi + \psi^* \nabla\psi \qquad (2.59)$$

so that

$$\text{Im}\{\mathbf{r}\psi^* \nabla\psi\} = \text{Im}\{\sum_{i=1}^{3} \partial/\partial r_i (\mathbf{r}\psi \partial/\partial r_i \psi)\} - \{\text{Im}\, \psi^* \nabla\psi\}$$

since the first term on the right hand side of (2.59) is real. Thus in (2.58)

$$\frac{d}{dt}\langle\mathbf{r}\rangle = \left(\frac{\hbar}{m}\right) \text{Im} \left\{ \int d^3\mathbf{r} \sum_{i=1}^{3} \partial/\partial r_i (\mathbf{r}\psi^* \partial/\partial r_i \psi) - \int d^3\mathbf{r}\, \psi^* \nabla\psi \right\} \qquad (2.60)$$

The first term on the right-hand side of (2.60) is zero by the divergence theorem (since it is equal to a surface term which can be made as small as we like by taking the surface as far away as we like), and the second term in the square brackets is purely imaginary since, integrating by parts and again dropping the surface term,

$$\left(\int d^3\mathbf{r}\, \psi^* \nabla\psi\right)^* = \int (\nabla\psi^*)\psi \, d^3\mathbf{r} = -\int d^3\mathbf{r}\, \psi^* \nabla\psi$$

Thus in (2.60)

$$(d/dt)\langle\mathbf{r}\rangle = (\hbar/mi) \int \psi^* \nabla\psi \, d^3\mathbf{r} = \langle\mathbf{p}\rangle/m$$

as required, with $\langle\mathbf{p}\rangle = (\hbar/i) \int \psi^* \nabla\psi \, d^3\mathbf{r}$. Similarly

$$\frac{d}{dt}\langle\mathbf{p}\rangle = \frac{\hbar}{i} \int \left\{ \frac{\partial \psi^*}{\partial t} \nabla\psi + \psi^* \nabla \frac{\partial \psi}{\partial t} \right\} d^3\mathbf{r}$$

$$= \frac{\hbar}{i} \int d^3\mathbf{r} \left[-\left(\frac{1}{i\hbar}\right) \left\{ -\frac{\hbar^2}{2m} \nabla^2\psi^* + V\psi^* \right\} \nabla\psi \right.$$

$$\left. + \frac{\psi^*}{i\hbar} \nabla\left\{ -\frac{\hbar^2 \nabla^2}{2m} \psi + V\psi \right\} \right]$$

$$= \int d^3\mathbf{r} \{V\psi^* \nabla\psi - \psi^* \nabla(V\psi)\}$$

$$- \frac{\hbar^2}{2m} \int d^3\mathbf{r} \{(\nabla^2\psi^*) \nabla\psi - \psi^* \nabla(\nabla^2\psi)\}$$

$$= -\int \psi^*\psi \nabla V \, d^3\mathbf{r} - \frac{\hbar^2}{2m}\int d^3\mathbf{r} \sum_{i=1}^{3} \partial/\partial r_i \{\partial/r\partial_i\psi^* \nabla\psi - \psi^* \partial/\partial r_i \nabla\psi\}$$

$$= -\langle \nabla V \rangle \tag{2.61}$$

as required, since the second term in (2.61) is zero by the divergence theorem.

We see that the average position and momenta of a state described by a wave function move like a classical particle. We will discuss in more detail in the next chapter under what conditions the *total* motion of the state is comparable to that of a classical particle.

PROBLEMS

2.1 On the basis of the Sommerfeld-Wilson quantisation rule, discuss the energy levels for

(a) A rigid rotator in 3 dimensions, rotating about a fixed axis, and derive a selection rule by means of the correspondence principle.

(b) The anharmonic oscillator in 1 dimension, with classical energy
$$E = p^2/2m + \tfrac{1}{2}kx^2 + \tfrac{1}{4}\lambda x^4 + \tfrac{1}{5}\mu x^5$$
(m, k, λ, μ are constants).

(c) The three dimensional harmonic oscillator, and discuss the degeneracy of the levels (i.e., the number of levels with the same energy).

2.2 Use the Sommerfeld-Wilson quantisation rule to show that for any elliptic orbit in a hydrogen atom, the energy levels are $E_n = me^4/2(4\pi\epsilon_0 nh)^2$, where $n = n_r + n_\phi$, n_r and n_ϕ being integers with $J_\phi = n_\phi \hbar$, $J_r = n_r \hbar$, and (r,ϕ) being the polar co-ordinates of the electron with respect to the proton.

Determine the resulting degeneracy of the energy levels.

2.3 Determine the effect of taking a finite nuclear mass (instead of an infinitely heavy nucleus) on the energy levels of the hydrogen atom, in the Bohr-Rutherford model.

2.4 At what velocity is the de Broglie wave length of a neutron equal to that of a 1 keV X-ray?

2.5 What energy must an alpha particle (four times the weight of a proton) possess to give non-Coulombic scattering by a nucleus of radius 8×10^{-15} metres, assuming that the potential is strictly Coulombic outside the nucleus, but departs from this form within the nucleus?

2.6 Normalize the momentum space wave function $a(p) = Ne^{-\alpha|p|/\hbar}$, and show that the corresponding position space wave function is

$$\psi(\mathbf{r}) = 1/\pi(2\alpha)^{3/2} \frac{\alpha}{(r^2+\alpha^2)^2}$$

Calculate Δx and Δp_x for this wave function, and so evaluate $\Delta x \, \Delta p_x$.

2.7 Calculate the probability current corresponding to the wave function

$$\psi = e^{ikr}/r$$

where $r = |\mathbf{r}| = (x^2+y^2+z^2)^{1/2}$. In particular interpret the form of this current for large values of r.

2.8 Show that for a normalized wave function, the associated probability current $\mathbf{S}$ is related to the average value $\langle \mathbf{p} \rangle$ of the momentum for the particle described by that wave function by

$$\int \mathbf{S} \, d^3\mathbf{r} = \langle \mathbf{p} \rangle / m$$

where m is the mass of the particle.
What is the physical interpretation of this relation?

FURTHER READING

1. BORN, M., *Atomic Physics*, Blackie, 1951; a good discussion of the experimental foundations of quantum mechanics.

2. TOMONAGA, S., *Quantum Mechanics*, vol. 1 and 2, North Holland, 1966; especially for the discussion of the difficulties of Classical mechanics and old quantum mechanics in vol. 1.

3. HEISENBERG, W., *Physical Principles of the Quantum Theory*, Dover Publ. 1930; a good qualitative discussion of the foundations of quantum mechanics.

4. DIRAC, P. A. M., *Quantum Mechanics*, Clarendon Press, 1957; a difficult but fundamental book.

CHAPTER 3

Motion in One Dimension

3.1 Introduction

At first sight, motion in one dimension might not appear to be of much use in helping us to understand bound states and scattering processes in our real space of three dimensions. This is in fact not the case. Indeed, a large class of three dimensional problems in quantum mechanics can be reduced to one dimensional motion in the distance $r = \sqrt{(x^2+y^2+z^2)}$ from the centre of force, since the angular variation of the wave function is relatively trivial and does not depend on the potential. Such problems are those for which the potential is symmetric about the origin, so is a function of r only, and the force is a central one. We will discuss such potentials in more detail in the next chapter; our understanding gained from considering one dimensional motion will be of great help in that context.

A further point in favour of considering one dimensional motion is that certain physical problems in three dimensions do actually reduce to one dimensional ones. Thus a particle moving in three dimensional space under the influence of identical central forces from a plane of scattering centres reduces to a problem of motion in one dimension, this dimension being the distance of the particle from the plane. Such a model has recently proved useful in considering particle motion in crystals, and is related to the phenomenon of chanelling; we will consider this later in the chapter.

Both the time dependent and the time independent Schrödinger equations (2.6) and (2.6′) are impossible to solve explicitly for nearly all potentials. This is especially so in three dimensions, partly due to the number of variables in the equations. If we consider the motion of a particle along a line, which is in one dimension, we have reduced the number of variables to one for the time independent Schrödinger equation (2.6) and to two for the time dependent Schrödinger equation (2.6′); such a reduction will enable us to solve these equations for various potentials. In this way we may set up exact solutions to the Schrödinger equation in three dimensions for a certain class of central force problems.

We will only consider in this chapter the time independent Schrödinger equation (2.6); thus we are only attempting to find the stationary states of certain one dimensional systems. The Schrödinger equation is thus:

$$-\frac{\hbar^2}{2m}\frac{d^2\psi}{dx^2} + V(x)\psi = E\psi(x)$$

with

$$\psi(x,t) = e^{-iEt/\hbar}\psi(x)$$

being the time dependent wave function.

We write the equation as

$$\psi'' + \frac{2m}{\hbar^2}[E - V(x)]\psi = 0 \qquad (3.1)$$

where $\psi'' = d^2\psi/dx^2$.

3.2 The Step Potential

The simplest non-trivial potential in one dimension is the step potential shown in Fig. 3.1, which has functional form

$$\left.\begin{array}{ll} V(x) = 0 & (x < 0) \\ = V_0 & (x \geq 0) \end{array}\right\} \qquad (3.2)$$

and we may assume $V_0 > 0$ without loss of generality. The Schrödinger equation (3.1) is thus

$$\left.\begin{array}{ll} \psi'' + (2mE/\hbar^2)\psi = 0 & (x < 0) \\ \psi'' + \{2m(E - V_0)/\hbar^2\}\psi = 0 & (x \geq 0) \end{array}\right\} \qquad (3.3)$$

MOTION IN ONE DIMENSION

If we introduce k_0 and k so that $2mE/\hbar^2 = k_0^2$, $2m(E-V_0)/\hbar^2 = k^2$, then (3.3) becomes

$$\left.\begin{array}{l}\psi'' + k_0^2 \psi = 0 \quad (x < 0) \\ \psi'' + k^2 \psi = 0 \quad (x \geqslant 0)\end{array}\right\} \quad (3.4)$$

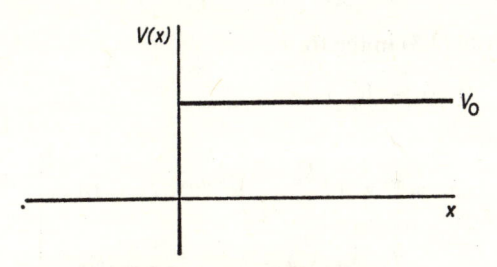

Fig. 3.1 The step potential.

The general solution of the two equations in (3.4) is a sum of plane waves:

$$\left.\begin{array}{rl}\psi(x) = & Ae^{ik_0 x} + Be^{-ik_0 x} \quad (x < 0) \\ = & Ce^{ikx} + De^{-ikx} \quad (x \geqslant 0)\end{array}\right\} \quad (3.5)$$

In order that the Schrödinger equation (3.1) be valid for *all* x, including $x = 0$, it is necessary that ψ and ψ' are continuous everywhere, *especially* at $x = 0$. For the solution represented by (3.5) to satisfy these continuity conditions, we must have ψ and ψ' continuous at $x = 0$ giving respectively:

$$\left.\begin{array}{l}A + B = C + D \\ ik_0 A - ik_0 B = ikC - ikD\end{array}\right\} \quad (3.6)$$

We may eliminate B and C, say, from (3.6) by substituting

$$\left.\begin{array}{l}B = \left(\dfrac{k_0 - k}{k_0 + k}\right) A + \left(\dfrac{2k}{k_0 + k}\right) D \\ C = \left(\dfrac{2k_0}{k_0 + k}\right) A + \left(\dfrac{k - k_0}{k + k_0}\right) D\end{array}\right\} \quad (3.7)$$

Hence

$$\begin{aligned}\psi(x) &= A\left(e^{ik_0 x}+\frac{k_0-k}{k_0+k}e^{-ik_0 x}\right)+D\frac{2k}{k_0+k}e^{-ik_0 x} \quad (x<0)\\ &= A\left(\frac{2k_0}{k+k_0}\right)e^{ikx}+D\left(\frac{k-k_0}{k+k_0}e^{ikx}+e^{-ikx}\right) \quad (x\geq 0)\end{aligned} \quad (3.8)$$

We can rewrite (3.8) in the form

$$\psi = A\psi_1 + D\psi_2 \tag{3.9}$$

where

$$\begin{aligned}\psi_1 &= e^{ik_0 x}+\left(\frac{k_0-k}{k_0+k}\right)e^{-ik_0 x} \quad (x<0)\\ &= \left(\frac{2k_0}{k_0+k}\right)e^{ikx} \quad (x\geq 0)\\ \psi_2 &= \left(\frac{2k}{k+k_0}\right)e^{-ik_0 x} \quad (x<0)\\ &= \left(\frac{k-k_0}{k+k_0}\right)e^{ikx}+e^{-ikx} \quad (x\geq 0)\end{aligned} \quad (3.10)$$

The waves ψ_1 and ψ_2 each have a simple physical interpretation which we shall now obtain. First of all we notice that the wave $e^{ik_0 x}$ has time dependent form $e^{ik_0 x-(iEt/\hbar)}$; it will represent a plane wave moving to the right, since as t increases x must also increase to keep the same phase $(k_0 x - Et/\hbar)$ of the wave. Similarly $e^{-ik_0 x-(iEt/\hbar)}$ represents a plane wave moving to the left. Thus ψ_1 represents the sum of a plane wave moving to the right towards the step of the potential at $x=0$ plus a wave moving to the left from this step, which may be regarded as a *reflection* of that right-moving wave by the potential step at $x=0$; for $x>0$, ψ_1 is a wave moving only to the right. It follows that ψ_1 represents an incoming wave $e^{ik_0 x}$ moving to the right towards the step at $x=0$ plus a reflected wave $Re^{-ik_0 x}$ reflected back from the step, where the *reflection coefficient* R is given by

$$R = (k_0-k)/(k_0+k)$$

There is also a part of ψ_1 which is *transmitted* through the step and

MOTION IN ONE DIMENSION

has the form Te^{ikx}, where the *transmission coefficient* T is given by

$$T = 2k_0/(k_0+k).$$

We note especially that T and R are defined in terms of an incoming wave part of ψ_1 with unit amplitude, i.e., just $e^{ik_0 x}$, so that T and R measure the amplitude of the transmitted and reflected wave relative to that of the incoming wave. A pictorial representation of this reflection and transmission process by the step, is shown in Fig. 3.2. We may evidently give a similar interpretation for ψ_2; from now on we just consider ψ_1.

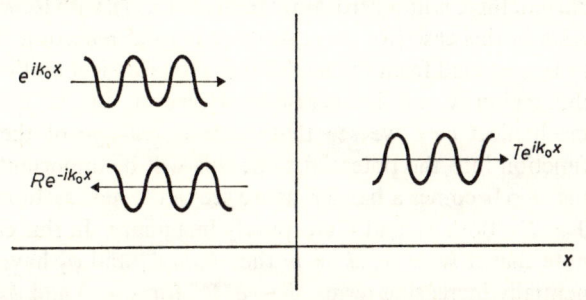

Fig. 3.2 The process of reflection and transmission of the incoming wave $e^{ik_0 x}$ by the step potential. The reflected wave moving to the left has amplitude equal to the reflection coefficient R; the transmitted wave has amplitude equal to the transmission coefficient T.

Whilst the above interpretation is in general true for all possible values of the energy E for a fixed height V_0 of the step, there are further differences brought about by different possibilities for E. There are in fact three different ranges of E:

(i) $E > V_0$ Then k is real, as is k_0, of course, so that ψ_1 contains truly oscillating waves and the above interpretation is exactly valid (with a similar interpretation for ψ_2). In general the probability current is, for $E \geqslant 0$,

$$\left. \begin{aligned} S &= \frac{\hbar}{m} \operatorname{Im}\left(\psi_1^* \frac{d\psi_1}{dx}\right) \\ &= \frac{\hbar k_0}{m}(1-|R|^2) \qquad (x < 0) \\ &= \frac{\hbar}{m}|T|^2 \operatorname{Re}(k) e^{-2x\operatorname{Im}(k)} \qquad (x \geqslant 0) \end{aligned} \right\} \qquad (3.11)$$

F

where in this case k is real. In the corresponding classical motion all values of x are accessible to the particle.

(ii) $0 < E < V_0$ k_0 is real, but k is purely imaginary, $k = i\kappa$, with $\kappa > 0$. Then for $x > 0$, ψ_1 is an exponentially damped function

$$\psi_1 = \left(\frac{2k_0}{k_0 + i\kappa}\right) e^{-\kappa x} \qquad (3.12)$$

The penetration into the step is to a distance of order $(1/\kappa)$, there is no actual *flow* into the step, since the probability current there is now zero, as we see from (3.11), with $\text{Re}(k) = 0$. Also in this case $|R|^2 = 1$, so there is total reflection. This is to be expected from the fact that for the classical motion only the region $x < 0$ is accessible. However, in the quantum mechanical case we see that there is *leakage* of the wave function into the potential step; this will be important when the step becomes a barrier, as we see in the next section.

(iii) $0 > E$ Both k_0 and k are purely imaginary. In this case we note that if $k_0 = i\kappa_0$, $k = i\kappa$ then *both* ψ_1 and ψ_2 have exponentially increasing terms, $\psi \sim e^{-\kappa_0 x}$ for $x < 0$ and $\psi_2 \sim e^{\kappa x}$ for $x > 0$ (κ_0, κ both taken as positive). Either of these exponentially increasing wave functions has to be rejected as physically meaningful, since the probability interpretation implies that for both of them the particle will have probability one of being at infinite values of x. There is thus no satisfactory quantum mechanical solution in this case; this corresponds to the fact that there is no region accessible to the classical motion.

3.3 The Potential Barrier

This potential is shown in Fig. 3.3; it is given by

$$\left. \begin{array}{ll} V(x) = 0 & (x \leqslant 0 \text{ or } x \geqslant a) \\ = V_0 & (0 < x < a) \end{array} \right\} \qquad (3.13)$$

with $V_0 \geqslant 0$.

The Schrödinger equation is very similar to (3.3), and takes the form

MOTION IN ONE DIMENSION

$$\psi'' + (2mE/\hbar^2)\psi = 0 \quad (x \leqslant 0 \text{ or } x \geqslant a)$$
$$\psi'' + (2m(E-V_0)/\hbar^2)\psi = 0 \quad (0 < x < a) \quad (3.14)$$

Let us define $2mE/\hbar^2 = k_0^2$, $2m(E-V_0)/\hbar^2 = k^2$, as in §3.2. Then (3.14) becomes

$$\psi'' + k_0^2 \psi = 0 \quad (x \leqslant 0 \text{ or } x \geqslant a)$$
$$\psi'' + k^2 \psi = 0 \quad (0 < x < a) \quad (3.15)$$

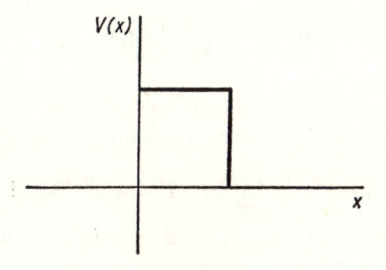

Fig. 3.3 The potential barrier.

We restrict our discussion to the case $E > 0$, and consider a wave incident on the barrier from the left. This means that in the region $x \geqslant a$ there will only be a wave travelling to the right. The form of ψ satisfying (3.15) of this nature will then be

$$\psi = e^{ik_0 x} + R e^{-ik_0 x} \quad (x \leqslant 0)$$
$$= A e^{ikx} + B e^{-ikx} \quad (0 < x < a) \quad (3.16)$$
$$= T e^{ik_0 x} \quad (x \geqslant a)$$

We see that R and T will be the reflection and transmission coefficients respectively, as defined in the last section. If we now apply the conditions that ψ and ψ' are continuous at $x = 0$ and at $x = a$, we have that

$$1 + R = A + B$$
$$k_0(1 - R) = k(A - B)$$
$$T e^{ik_0 a} = A e^{ika} + B e^{-ika} \quad (3.17)$$
$$k_0 T e^{ik_0 a} = k(A e^{ika} - B e^{-ika})$$

We may solve (3.17) for A, B, R and T by elimination: the first and second of the equations give

$$\left.\begin{aligned} A &= \frac{1}{2}(1+R)+\frac{1}{2}\frac{k_0}{k}(1-R) \\ B &= \frac{1}{2}(1+R)-\frac{1}{2}\frac{k_0}{k}(1-R) \end{aligned}\right\} \qquad (3.18)$$

whilst the third and fourth equations, eliminating T give

$$Ae^{ika}+Be^{-ika} = \left(\frac{k}{k_0}\right)(Ae^{ika}-Be^{-ika}) \qquad (3.19)$$

(3.18) and (3.19) may be used to obtain R, and (3.17) then used to obtain T; the result of this, after a little simplification, is

$$\left.\begin{aligned} R &= (1-\mu^2)\sin ka/D, \quad T = e^{-ik_0 a}2i\mu/D \\ A &= e^{-ika}i(1+\mu)/D, \quad B = -ie^{ika}(1-\mu)/D \end{aligned}\right\} \qquad (3.20)$$

where $D = (1+\mu^2)\sin ka + 2i\mu\cos ka$, $\mu = k/k_0$. These formulae are still correct, even when $E < V_0$, so that k is purely imaginary. In both cases $E < V_0$ or $E > V_0$, we have $|T|^2 + |R|^2 = 1$; this is always to be expected since *either* reflection *or* transmission of the wave must occur at the barrier. However, the nature of the waves will be different in the two cases. When $E > V_0$ both k and k_0 are real; the incoming wave $e^{ik_0 x}$ in (3.16) either suffers reflection at $x = 0$, so as to be the wave $Re^{-ik_0 x}$ travelling to the left, or continues *as a wave motion* to the right of $x = 0$ with form $Ae^{ikx}+Be^{-ikx}$. The coefficients A and B have to be chosen so that only a transmitted wave $Te^{ik_0 x}$ moving to the right emerges from the barrier at $x = a$. This is similar to the nature of the classical motion, for which all values of x are accessible. In particular there is complete transmission, $|T| = 1$, when $ka = n\pi$, so there are an integral number of waves in the barrier. When $E < V_0$ the particle *cannot* move classically into the region $0 < x < a$, so the barrier is impenetrable. The quantum mechanical motion is remarkably different from this and there is *leakage* of the particle through the barrier. In this case we can deduce from (3.20), that when $k = i\kappa$ and $\mu = i\lambda$:

$$|T|^2 = (4\lambda)^2/\{(1-\lambda^2)^2\sinh^2\kappa a + 4\lambda^2\cosh^2\kappa a\} \qquad (3.21)$$

If the potential barrier is high compared to the energy of the particle, so that

$$\kappa^2 \sim 2mV_0/\hbar^2, \quad \lambda^2 \sim V_0/E \gg 1,$$

then (3.21) becomes

$$|T|^2 \sim 16/(\lambda^2 \sinh^2 \kappa a)$$

If further the barrier is broad, so that $\kappa a \gg 1$, then this becomes

$$|T|^2 \sim (16E/V_0)e^{-2\kappa a} \tag{3.22}$$

The result of (3.22) is that the transmission probability is much less than one, and decreases exponentially with increasing width of barrier. Such leakage of a particle through a barrier was first used to explain the α-decay of the nucleus by Gamow in 1928 when he assumed that the α-particle was held in the nucleus by a potential as shown in Fig. 3.4; the shape of the wave function in such a case is also shown in Fig. 3.4.

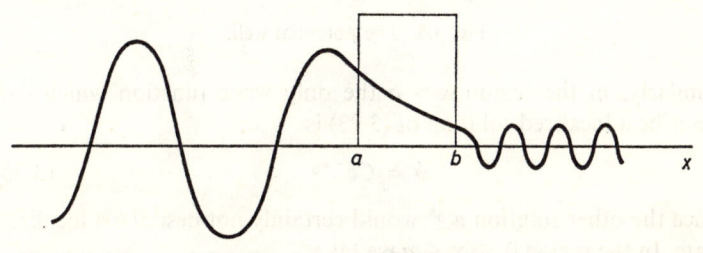

Fig. 3.4 The wave function in a potential barrier similar to that in a nucleus.

3.4 The Potential Well

This example is extremely important as a model which shows how a discrete set of energies is allowed when solving the Schrödinger equation (2.6) for bound (localized) states. The form of the potential is shown in Fig. 3.5; its analytic form is as given in (3.13), where now $V_0 < 0$. The Schrödinger equation still takes the form (3.15). We are interested in the situation which classically would correspond to the particle trapped in the region $0 \leqslant x \leqslant a$, for which $E < 0$. In that case $k_0 = i\kappa_0$ where $\kappa_0 > 0$, and the only decreasing *localized* wave satisfying the Schrödinger equation for $x < 0$:

$$\psi'' - \kappa_0^2 \psi = 0 \tag{3.23}$$

will be $e^{\kappa_0 x}$; $e^{-\kappa_0 x}$ would *increase* exponentially as $x \to -\infty$, so could not describe a localized state. Thus we take

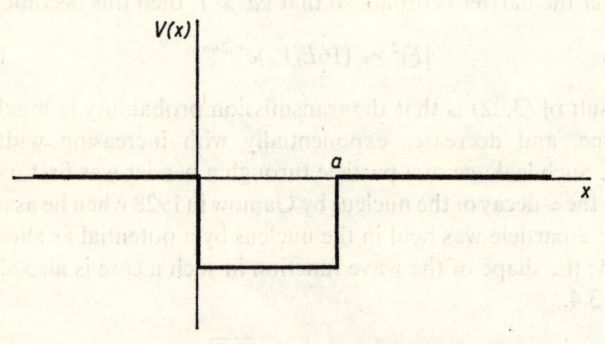

$$\psi = e^{\kappa_0 x} \tag{3.24}$$

Fig. 3.5 The potential well.

Similarly, in the region $x > a$ the only wave function which can describe a localized solution of (3.23) is

$$\psi = Ce^{-\kappa_0 x} \tag{3.25}$$

since the other solution $e^{\kappa_0 x}$ would certainly not describe a localized state. In the region $0 \leqslant x \leqslant a$ we take

$$\psi = Ae^{ikx} + Be^{-ikx}$$

as in (3.16), where k will be real if $E > V_0$ and will be purely imaginary if $E < V_0$. The requirement of continuity of ψ and ψ' at $x = 0$ and $x = a$ will give

$$\left.\begin{aligned} 1 &= A+B \\ \kappa_0 &= ik(A-B) \\ Ce^{-\kappa_0 a} &= Ae^{ika} + Be^{-ika} \\ -\kappa_0 Ce^{-\kappa_0 a} &= ik(Ae^{ika} - Be^{-ika}) \end{aligned}\right\} \tag{3.26}$$

We see that (3.26) consists of four equations for three unknowns A, B, C; there will thus be a consistency condition relating κ_0 and k.

It is this consistency condition which will determine E in terms of V_0. To see this we solve the first two equations of (3.26) to give:

$$A = \tfrac{1}{2}(1 - i\kappa_0/k), \quad B = \tfrac{1}{2}(1 + i\kappa_0/k) \quad (3.27)$$

We then find from the third equation of (3.26) that

$$\begin{aligned} C &= \tfrac{1}{2}e^{\kappa_0 a}\left\{\left(1 - i\frac{\kappa_0}{k}\right)e^{ika} + \left(1 + i\frac{\kappa_0}{k}\right)e^{-ika}\right\} \\ &= e^{\kappa_0 a}\left(\cos ka + \frac{\kappa_0}{k}\sin ka\right) \end{aligned} \quad (3.28)$$

If we now insert the values of A, B and C given by (3.27) and (3.28) into the last equation of (3.26) we obtain

$$\begin{aligned} -\kappa_0\left\{\cos ka + \frac{\kappa_0}{k}\sin ka\right\} &= \left(i\frac{k}{2}\right)\left\{e^{ika}\left(1 - i\frac{\kappa_0}{k}\right) - e^{-ika}\left(1 + i\frac{\kappa_0}{k}\right)\right\} \\ &= ik\left\{i\sin ka - i\frac{\kappa_0}{k}\cos ka\right\} \end{aligned}$$

or

$$-2\kappa_0 \cos ka = \left(-k + \frac{\kappa_0^2}{k}\right)\sin ka$$

or finally

$$2 \cot ka = \frac{k}{\kappa_0} - \frac{\kappa_0}{k} \quad (3.29)$$

This is the desired relation between E and V_0; we may simplify it if we introduce new parameters

$$\gamma = (-2mV_0 a^2/\hbar^2)^{\frac{1}{2}}, \quad \alpha = ka/\gamma = \{1 - (E/V_0)\}^{\frac{1}{2}} \quad (3.30)$$

Then $0 < \alpha < 1$ when $V_0 < E < 0$, and

$$ka = \alpha\gamma, \quad (k^2 - \kappa_0^2)/k\kappa_0 = (2\alpha^2 - 1)/\alpha\sqrt{(1-\alpha^2)}$$

so that (3.29) becomes

$$2 \cot \alpha\gamma = (2\alpha^2 - 1)/\alpha\sqrt{(1-\alpha^2)}$$

But this implies

$$\cos \alpha\gamma = 2\alpha^2 - 1$$

or

$$\tfrac{1}{2}\alpha\gamma = \cos^{-1}\alpha + \tfrac{1}{2}n\pi \quad (n = 0, 1, 2, \cdots) \quad (3.31)$$

i.e.
$$\{2m(E-V_0)a^2/\hbar^2\}^{\frac{1}{2}} = 2\cos^{-1}\{1-E/V_0\}^{\frac{1}{2}} + n\pi$$

We solve (3.31) by plotting both the left hand side, $\frac{1}{2}\alpha\gamma$, and the right hand side, $\cos^{-1}\alpha + \frac{1}{2}n\pi$, as functions of α; their graphs are shown in Fig. 3.6; the graph of $\frac{1}{2}\alpha\gamma$ is plotted for three different values of

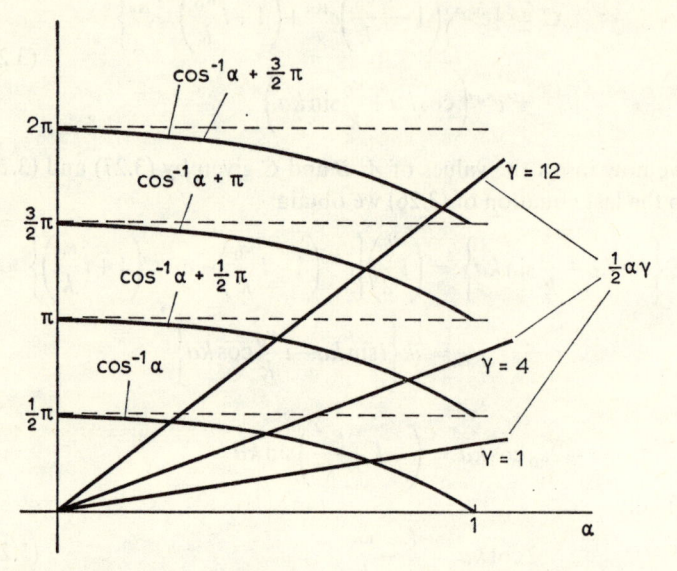

Fig. 3.6 The graphs of $\frac{1}{2}\alpha\gamma$ for $\gamma = 1, 4$ and 12, and of $\cos^{-1}\alpha + \frac{1}{2}n\pi$ for $n = 0, 1$ and 2, plotted against α.

$\gamma, \gamma = 1, 4$, and 12. The curve $\gamma = 1$ intersects only the curve $\cos^{-1}\alpha$, so there is only one value of α satisfying (3.31), that for $n = 0$. We see that as γ increases to π the curve $\frac{1}{2}\alpha\gamma$ just meets the next higher curve $\cos^{-1}\alpha + \frac{1}{2}\pi$, as well as the curve $\cos^{-1}\alpha$. The meeting with the curve $\cos^{-1}\alpha + \frac{1}{2}\pi$ is at $\alpha = 1$ when $\gamma = \pi$; as γ increases above π the common point on $\cos^{-1}\alpha + \frac{1}{2}\pi$ and $\frac{1}{2}\alpha\gamma$ has its value of α decreasing; this means that as γ increases through π a localized bound state is allowed with energy $E = 0$ (when $\alpha = 1$ at $\gamma = \pi$). The energy of this bound state becomes more and more negative as γ increases above π. A further bound state appears with zero energy when γ passes through 2π, and similarly when γ passes through $3\pi, 4\pi, \cdots$. We note that even for arbitrarily small γ there is always one bound state, whose energy tends to zero as γ vanishes. We may understand this situation physically as related to what happens as the depth $|V_0|$ or width a of the well are increased; this increase pulls the particle more and more strongly into the well.

The form of the wave functions for these bound states is very instructive; it will enable us to understand more clearly why only certain discrete energies are allowed for the bound states. The wave function may be written down from the earlier discussion in this section; in particular if we use (3.27) we have that for $0 < x < a$

$$\psi(x) = \cos kx + (\kappa_0/k)\sin kx \tag{3.32}$$

We may replace the factor (κ_0/k) by $\sqrt{(1-\alpha^2)}/\alpha$, so by $\pm \tan\tfrac{1}{2}ka$, giving in (3.32):

$$\psi(x) = \pm(\cos\tfrac{1}{2}ka)^{-1}\{\cos kx \cos\tfrac{1}{2}ka + \sin kx \sin\tfrac{1}{2}ka\}$$

$$= \frac{1}{\alpha}\cos\{k(x - a/2) + \tfrac{1}{2}n\pi\} \tag{3.33}$$

The n entering in (3.33) is the same as that in (3.31) and takes account of the multiplicity of solutions in exactly the same fashion. In a similar fashion we see that C given by (3.28) is obtained by taking $x = a$ in (3.32), so has the value $C = (-1)^n e^{\kappa_0 x}$; we thus obtain

$$\left.\begin{array}{ll}\psi = e^{\kappa_0 x} & (x < 0) \\ = \dfrac{1}{\alpha}\cos\left\{k\left(x - \dfrac{a}{2}\right) + \tfrac{1}{2}n\pi\right\} & (0 < x < a) \\ = (-1)^n e^{-\kappa_0(x-a)} & (a < x)\end{array}\right\} \tag{3.34}$$

We see that ψ vanishes at exactly n points between 0 and a and nowhere else; the shape of ψ is shown for $n = 0, 1, 2$ in Fig. 3.7. The wave function is limited by the four conditions for suitably joining it up to exponentially damped wave functions at $x = 0$ and $x = a$; it can do so only with one arbitrary quantity, and that is the number n of its zeros. The lower n is, the less the oscillatory character of the wave function, so the less kinetic energy of the state, and the easier the state is 'trapped' as a bound state. As n increases, so do the number of oscillations of the wave function and so does the kinetic energy; to bind such a wave function, the potential must become stronger by being deeper or wider. This gives, then, an explanation of the process of the formation of a discrete set of bound states in a potential well, and shows how the behaviour of the wave function is related to the energy level of the state.

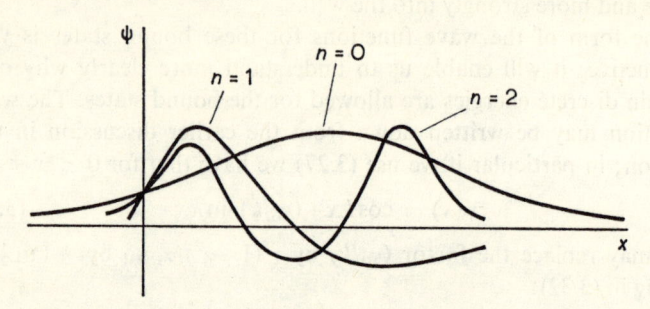

Fig. 3.7 The bound state wave function in the potential well for the three lowest energy levels, with 0, 1 and 2 finite nodes.

To conclude this section we let $V_0 \to -\infty$, so that the particle is in a one-dimensional box of length a. In that case the graph of $\frac{1}{2}\alpha\gamma$ becomes a vertical line from $\alpha = 0$, and the points of intersection with $\cos^{-1}\alpha + \frac{1}{2}\pi n$ are at $\alpha\gamma = n\pi$. Dropping the common potential energy, the energies are

$$E_n = \left(\frac{\hbar^2 k^2}{2m} + V_0\right) - V_0 = \frac{n^2\pi^2\hbar^2}{2ma^2}$$

This is precisely the value given at the end of §2.2 by the Sommerfeld-Wilson quantisation rules.

3.5 The Harmonic Oscillator

Another important example with many physical applications is the harmonic oscillator, for which the classical potential is $V(x) = \frac{1}{2}kx^2$ discussed in §1.2. For a particle of mass m and momentum p the classical Hamiltonian is

$$H = (p^2/2m) + \frac{1}{2}kx^2 \qquad (3.35)$$

whilst the classical equations of motion which follow from this are

$$p = m\dot{x}, \qquad \dot{p} = -kx$$

MOTION IN ONE DIMENSION

so that
$$m\ddot{x} + kx = 0 \qquad (3.36)$$
The resulting motion is simple harmonic of general form
$$x = A\sin(\omega t + \theta) \qquad (3.37)$$
where $\omega^2 = k/m$, and the period of the classical motion is $T = 2\pi/\omega$; the limits of the classical motion are at the values of x for which p is zero, so satisfy
$$E = \tfrac{1}{2}kx^2$$
The Schrödinger equation arising from (3.35) is
$$\psi'' + (2m/\hbar^2)(E - \tfrac{1}{2}kx^2)\psi = 0 \qquad (3.38)$$
We remove as many free parameters as possible by introducing the new variable y with $x = (\hbar/m\omega)^{\frac{1}{2}}y$ and define $E = \lambda\hbar\omega$, so that (3.38) becomes
$$\frac{d^2\psi}{dy^2} + (2\lambda - y^2)\psi = 0 \qquad (3.39)$$
where ψ is regarded now as a function of y, $\psi = \psi(y)$.

In order to solve the modified Schrödinger equation (3.39), we may proceed by means of annihilation and creation operators. These are the differential operators $[(d/dy) \pm y]$; let us see what these produce when they act on the wave function ψ (which we denote by ψ_λ to show its explicit dependence on λ) which satisfies (3.39). We have (denoting d/dy by D):
$$\left.\begin{array}{l}(D+y)(D-y) = D^2 - y^2 - 1 \\ (D-y)(D+y) = D^2 - y^2 + 1\end{array}\right\} \qquad (3.40)$$
so that
$$(D+y)(D-y)\psi_\lambda = (D^2 - y^2 - 1)\psi_\lambda$$
$$= (-1 - 2\lambda)\psi_\lambda \qquad (3.41)$$
$$(D-y)(D+y)\psi_\lambda = (1 - 2\lambda)\psi_\lambda \qquad (3.42)$$
Thus
$$(D-y)(D+y)(D-y)\psi_\lambda = (-2\lambda - 1)(D-y)\psi_\lambda \qquad (3.43)$$
and
$$(D+y)(D-y)(D+y)\psi_\lambda = (1 - 2\lambda)(D+y)\psi_\lambda \qquad (3.44)$$

From (3.43) we see that the function $(D-y)\psi_\lambda$ is a solution of the Schrödinger equation (3.39), though with λ replaced by $(\lambda+1)$, since (3.39) then becomes, using (3.41)

$$(D-y)(D+y)\psi_{\lambda+1} = (-1-2\lambda)\psi_{\lambda+1}$$

which is (3.43). Thus we can write

$$(D-y)\psi_\lambda = (\text{constant})\psi_{\lambda+1} \qquad (3.45)$$

where we are not interested in the actual value of this constant multiplying $\psi_{\lambda+1}$. The only case when this does not occur is when

$$(D-y)\psi_\lambda = 0$$

which has the solution

$$\psi_\lambda \propto e^{\frac{1}{2}y^2} \qquad (3.46)$$

Evidently the wave function (3.46) is highly *non-localized*, and can never describe a localized bound state. Hence $(D-y)\psi_\lambda \neq 0$ for any λ. In a similar fashion we may show that

$$(D+y)\psi_\lambda = (\text{constant})\psi_{\lambda-1} \qquad (3.47)$$

unless

$$(D+y)\psi_\lambda = 0, \quad \text{or} \quad \psi_\lambda \propto e^{-\frac{1}{2}y^2} \qquad (3.48)$$

In this case the wave function $e^{-\frac{1}{2}y^2}$ of (3.48) is a good candidate for a bound state wave function; for this, $\lambda = \frac{1}{2}$, since this is required by

$$0 = (D-y)(D+y)e^{-\frac{1}{2}y^2} = (1-2\lambda)e^{-\frac{1}{2}y^2}$$

(The last part of this equation arises from (3.42).) We see that if we start from some solution ψ_λ of (3.39) and apply $(D+y)$ repeatedly to this wave function, each application will reduce the value of λ by one for the resulting wave function, and will ultimately give as negative a value to the energy $E = \lambda\hbar\omega$ as we like. Arbitrarily negative energies are not allowed in quantum mechanics (or classical mechanics), so that repeated application of $(D+y)$ must ultimately produce the zero wave function. In other words $(D+y)$ applied enough times to *any* ψ_λ will ultimately produce the wave function $e^{-\frac{1}{2}y^2}$, which will then be annihilated by one further application of $(D+y)$. Thus $\psi_0 = e^{-\frac{1}{2}y^2}$ is the lowest energy state, with energy $\frac{1}{2}\hbar\omega$; the higher energy states are given by the wave functions

$$\psi_n = (D-y)^n\psi_0 = (D-y)^n e^{-\frac{1}{2}y^2} \qquad (3.49)$$

with energy $E_n = (n+\frac{1}{2})\hbar\omega$. We see that $(D-y)$ may be regarded as creating another quantum of energy $\hbar\omega$ when acting on ψ_n since $(D-y)\psi_n = \psi_{n+1}$; $(D-y)$ may thus be regarded as a creation operator of the elementary quantum of energy $\hbar\omega$ of the harmonic oscillator. In a similar fashion $(D+y)$ may be regarded as an annihilation operator of the elementary quantum $\hbar\omega$, with $(D+y)\psi_n = \psi_{n-1}$. We note that the commutation relation between $(D-y)$ and $(D+y)$ is

$$(D-y)(D+y) - (D+y)(D-y) = 2 \qquad (3.50)$$

which (except for a factor of $2i$) is the commutation relation for the momentum p and position q of a single particle; this is to be expected since $(D-y)$ and $(D+y)$ are linearly related to p and q:

$$D-y = (ip/\hbar) - q, \qquad D+y = (ip/\hbar) + q$$

We have explicit forms for the lower wave functions:

$$\psi_0(y) = e^{-\frac{1}{2}y^2}$$
$$\psi_1(y) = 2ye^{-\frac{1}{2}y^2}$$
$$\psi_2(y) = 2(2y^2 - 1)e^{-\frac{1}{2}y^2}$$

The general wave function ψ_n is the form of a polynomial times $e^{-\frac{1}{2}y^2}$; the polynomial is of degree n in y and called a *Hermite polynomial* $(H_n(y)$, so*

$$\psi_n(y) = H_n(y)e^{-\frac{1}{2}y^2}$$

The first three wave functions are shown in Fig. 3.8; we note that $\psi_n(y)$ has $(n+2)$ zeros (two of which are at $\pm\infty$, arising from the exponential damping term).

It is possible to discuss the Schrödinger equation (3.39) directly,

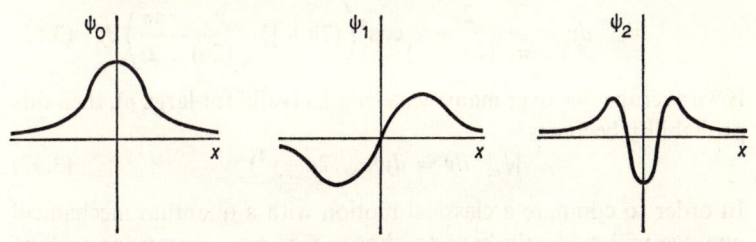

Fig. 3.8 The harmonic oscillator wave functions for the three lowest energy levels, with 0, 1 and 2 finite nodes.

*It is usual to define the Hermite polynomials so that $\psi_n(y)$ is normalised to unity, $\int_0^\infty [H_n(y)]^2 e^{-y^2} dy = 1$.

without use of the creation and annihilation operators; this may be done by considering the wave function $\phi(y) = e^{\frac{1}{2}y^2}\psi(y)$, and taking a power series expansion for ϕ. The coefficients of the succeeding powers of y in this expansion may be found from (3.39); unless they become zero after a finite number of terms, (so that ϕ is a polynomial) it may be shown that $\psi(y) \sim e^{\frac{1}{2}y^2}$ as $y \sim \infty$, so the state is not bound. The condition that ϕ is a polynomial is that $E = (n+\frac{1}{2})\hbar\omega$, (for some positive integer n or zero) which we found earlier; the resulting polynomial may be shown to be the Hermite polynomial $H_n(y)$ which we introduced earlier. We prefer to use creation and annihilation methods here since they are more related to the modern approach in physics.

We finally remark on the classical limit in the harmonic oscillator potential. Classically we have, with $\theta = 0$ in (3.37),

$$\frac{dx}{dt} = \omega A \cos \omega t$$

so the fraction of a whole period during which the particle is in the interval dx is

$$\frac{2dt}{T} = \frac{2dx}{T\omega A \cos \omega t} = \frac{2dx}{\omega T\sqrt{(A^2-x^2)}} = \frac{dx}{\pi\sqrt{(A^2-x^2)}}$$

Thus the classical probability $P(x)\,dx$ that the particle is in dx is

$$P(x) = 1/\pi\sqrt{(A^2-x^2)} \tag{3.51}$$

The quantum mechanical probability that the particle is in the interval dy may be shown to be, for large n,

$$|\psi_n|^2\,dy = \frac{2}{\pi\sqrt{(2n-y^2)}} \cos^2\left((2n+\tfrac{1}{2})\frac{y}{\sqrt{(2n)}} - \frac{n\pi}{2}\right) dy \tag{3.52}$$

If we average this over many wavelengths (valid for large n), then this probability becomes

$$|\psi_n|^2\,dy = dy/\pi\sqrt{(2n-y^2)} \tag{3.53}$$

In order to compare a classical motion with a quantum mechanical one, we will evidently have to choose the same energy for each of these motions. The quantum mechanical one has energy $n\hbar\omega$ (for large n), whilst the classical one has energy $\frac{1}{2}m\omega^2 A^2$, so that $A^2 = 2n\hbar/m\omega$; the classical probability of finding the particle in the

interval dy is then the value of $P(x)$ in (3.51) multiplied by (dx/dy), since
$$P(x)\,dx = P(x)\,(dx/dy)\,dy$$
Therefore
$$P(x)\frac{dx}{dy} = \left[1 \Big/ \left\{\pi\left(\frac{2n\hbar}{m\omega} - \frac{\hbar y^2}{m\omega}\right)^{\frac{1}{2}}\right\}\right]\left(\frac{\hbar}{m\omega}\right)^{\frac{1}{2}}$$
$$= 1/\{\pi\sqrt{(2n-y^2)}\}$$

which agrees with the averaged quantum mechanical probability distribution (3.53). Fig. 3.9 shows the quantum mechanical probability (3.52) and the classical probability (3.51); we see that the classical limit is obtained as $n \to \infty$, so that the wave length of the wave function tends to zero. Zero wave length implies that the wave cannot be diffracted round any obstacle; so this is indeed the classical picture!

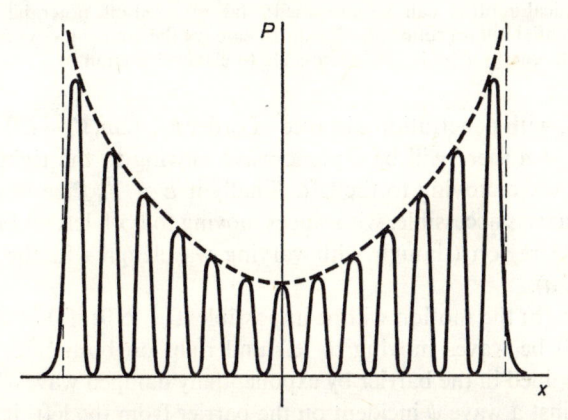

Fig. 3.9 The quantum mechanical probability distribution for a particle moving in a harmonic potential for large n, as shown by the continuous line; the dashed line denotes twice the classical probability distribution.

3.6 The General Potential

We can now make some remarks about the general nature of the solution to the Schrödinger equation for a general potential, using the understanding we have built up so far with the particular potentials dealt with in the previous five sections. We consider in particular three types of potential, shown in Fig. 3.10, (a), (b) and (c).

In case (a) we have that motion is impossible if $E < 0$ (as it is classically); if $0 < E < V_0$, motion will only be possible for some semi-infinite range bounded on the right by a, though there will be penetration into the potential step between 0 and a, depending on the

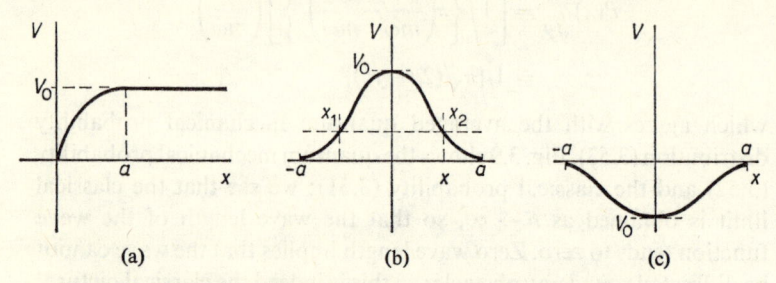

Fig. 3.10. Types of potential for which the general features of quantum mechanical motion can be discussed: (a) rounded-off potential step (b) potential hill (c) rounded-off well. In case (b) the segment $x_1 x_2$ of the line with energy $E(< V_0)$ is inaccessible to classical motion.

energy E, with penetration distance of order $\hbar/\sqrt{[2m(V_0-E)]}$; to the left of $x = a$ there will be a plane wave moving to the right and a reflected wave moving to the left. Finally if $E > V_0$ then the whole of the x-axis is accessible, with waves moving to both left and right in the whole region (though with varying wavelengths in the region $0 < x < a$).

In case (b) the motion will be impossible if $E < 0$; if $0 < E < V_0$ there will be waves moving to left and right on both sides of the barrier, joined in the barrier by exponentially damped waves. Thus if we take just a wave ψ incident on the barrier from the left, it will be partly reflected, with right-going wave $R\psi$ (for $x < -a$), and transmitted with transmitted wave $T\psi$ (for $x > a$). In the intermediate region where the wave is in the barrier, between the points x_1 and x_2 which are the limits of classical motion at the barrier, the wave is damped out exponentially as x increases, and we expect that T is of order $e^{-\kappa(x_2-x_1)}$, where again κ^{-1} is of order $\hbar/\sqrt{\{2m(V_0-E)\}}$.

In case (c) motion is impossible when $E < V_0$; when $E > 0$ the whole of the x-axis will be accessible and waves will be able to travel in both directions. For $0 > E > V_0$ we expect there only to be oscillatory waves in the classically accessible region for a discrete set of values of E: $E_0, E_1, E_2, \cdots$. The wave function for such a discrete

MOTION IN ONE DIMENSION

energy value will be exponentially damped out with distance into the classically inaccessible region; in the accessible region the wave function corresponding to the energy E_n will have n finite zeros (and 2 at $\pm \infty$) if the energies are ordered as $E_0 < E_1 < E_2 < \cdots$. We will turn to the effect of the potential on the states with $E > 0$ when we consider scattering in Chapter 6.

3.7 The Periodic Potential

In this section we will consider a very important one-dimensional potential, the periodic potential. This is a realistic potential to be used to describe the force field on a charged ion moving in a crystal lattice in suitable situations; the atoms of the crystal lattice may be regarded as lying on a series of parallel planes, and if the ion is moving fast nearly parallel to these planes, it will experience a force depending only on its perpendicular distance from the planes. Such a force will be describable by a periodic potential in one dimension, the variable x being the distance along the perpendicular to the planes, and the periodicity being the distance between two planes. The results we obtain for the properties of the solution for the one-dimensional periodic potential will also indicate the sort of properties we expect to hold for motion in a three-dimensional periodic potential; such a potential is certainly a valid description of the force field inside a crystal when it is regarded as a periodic lattice of atoms. We will see that the most important property of the wave function, that of *allowed and forbidden energy bands*, does indeed occur for the one-dimensional periodic potential.

We consider a general periodic potential $V(x)$ of the form shown in Fig. 3.11. The periodicity is a, and we consider explicitly the $(n-1)$st and nth intervals of periodicity; we suppose that there is an interval of constancy of the potential in each periodicity interval; the value of the potential in such an interval will be denoted by V_0. We take $\kappa = \{2m(E-V_0)/\hbar\}^{\frac{1}{2}}$, so that in each interval of constancy the Schrödinger equation will be

$$\psi'' + \kappa^2 \psi = 0 \qquad (3.54)$$

Hence the wave function will be a linear combination of plane waves travelling to the left and to the right:

$$\begin{aligned}\psi(x) &= l_{n-1}e^{-i\kappa x}+r_{n-1}e^{i\kappa x} \quad (x \text{ in } I_{n-1}) \\ &= l_n e^{-i\kappa x}+r_n e^{i\kappa x} \quad (x \text{ in } I_n)\end{aligned} \quad (3.55)$$

where $l_n, r_n, l_{n-1}, r_{n-1}$ are constants and I_{n-1}, I_n are the intervals in which the potential is constant for $(n-1)a \leq x \leq na$ and $na \leq x \leq (n+1)a$ respectively. We expect a linear relation between the constants l_{n-1}, r_{n-1} and l_n, r_n since the Schrödinger equation is a linear equation; there will be constants R and T so that

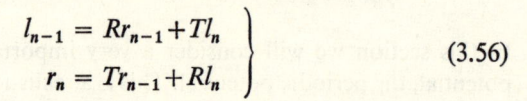

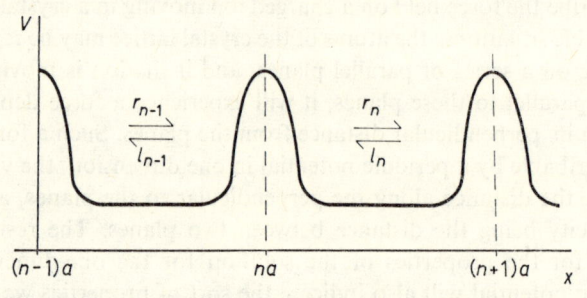

$$\begin{aligned} l_{n-1} &= Rr_{n-1}+Tl_n \\ r_n &= Tr_{n-1}+Rl_n \end{aligned} \quad (3.56)$$

Fig. 3.11 One-dimensional periodic potential used in describing motion in a crystal; the distance between two parallel planes of atoms in the crystal is a the amplitudes of waves travelling between the nth and $(n+1)$st planes to the left are r_n and l_n respectively.

We may interpret (3.56) in terms of reflection and transmission; the part of the wave function ψ in I_{n-1} moving to the left is composed of a contribution proportional to that moving to the right and then reflected from the potential barrier around $x = na$; the constant of proportionality R is naturally called the reflection coefficient. There is a further contribution arising from the transmitted part of the wave function in I_n moving to the left onto the potential barrier around $x = na$, with a proportionality constant T called the transmission coefficient. The second equation in (3.56) may be interpreted similarly as a sum of contributions to the part of the wave function in I_n moving to the right, coming from transmission through the potential barrier at $x = na$ of the wave function in I_{n-1} moving to the right plus the reflected part of the wave function in I_n moving to the left onto the barrier at $x = na$. The constants R and T are independent of n, and depend only on the potential V taken, say, for $0 \leqslant x \leqslant a$; they can be evaluated if an explicit form of V is given, though we will not discuss that here since we want to obtain results valid for a large class of potentials of the general form shown in Fig. 3.11.

We now consider the two wave functions $\psi(x)$ and $\psi(x+a)$; both of these satisfy the Schrödinger equation with energy E and potential $V(x)$ and $V(x+a)$ respectively:

$$\left\{-\frac{\hbar^2}{2m}\frac{d^2}{dx^2}+V(x)\right\}\psi(x) = E\psi(x) \qquad (3.57)$$

$$\left\{-\frac{\hbar^2}{2m}\frac{d^2}{dx^2}+V(x+a)\right\}\psi(x+a) = E\psi(x+a) \qquad (3.58)$$

But $V(x)$ is a periodic function in x with period a, $V(x) = V(x+a)$, so equations (3.57) and (3.58) imply that *both* $\psi(x)$ and $\psi(x+a)$ are eigenstates of the Hamiltonian $H = -(\hbar^2/2m)\,d^2/dx^2 + V(x)$ belonging to the eigenvalue E. If we assume that there is only *one* wave function (to within a constant) which has energy E for this Hamiltonian, then there is a complex number λ such that

$$\psi(x+a) = \lambda\psi(x) \qquad (3.59)$$

From this follows that

$$\psi(x \pm na) = \lambda^{\pm n}\psi(x) \qquad (3.60)$$

But since $|\psi(x \pm na)|$ is bounded as $n \to \infty$ we must have $|\lambda| = 1$, otherwise one or other of $\lim_{n\to\infty} |\psi(x \pm na)|$ will be unbounded. Thus there is a real number k for which

$$\psi(x+a) = e^{ika}\psi(x) \qquad (3.61)$$

This relation is known as Floquet's theorem. If we now use the forms in (3.55) in this relation, we obtain for x in the interval I_{n-1}

$$e^{ika}\{l_{n-1}e^{-i\kappa x}+r_{n-1}e^{i\kappa x}\} = l_n e^{-i\kappa(x+a)}+r_n e^{i\kappa(x+a)} \qquad (3.62)$$

Since this is true for x in a non-zero interval, the coefficients of the waves $e^{i\kappa x}$ and $e^{-i\kappa x}$ on both sides of this equation must be zero, so

$$\left.\begin{array}{l} l_n = e^{i(\kappa+k)a}l_{n-1} \\ r_n = e^{i(k-\kappa)a}r_{n-1} \end{array}\right\} \qquad (3.63)$$

If we now use (3.63) to substitute for l_{n-1} and r_n in terms of l_n and r_{n-1} in (3.56) we obtain

$$\begin{pmatrix} T - e^{-i(k+\kappa)a} & R \\ R & T - e^{i(k-\kappa)a} \end{pmatrix} \begin{pmatrix} l_n \\ r_{n-1} \end{pmatrix} = \begin{pmatrix} 0 \\ 0 \end{pmatrix} \quad (3.64)$$

The condition for a non-zero solution of this two dimensional vector equation is that the determinant of the 2×2 matrix on the left hand side of (3.64) is zero,

$$\{T - e^{-i(k+\kappa)a}\} \{T - e^{i(k-\kappa)a}\} = R^2 \quad (3.65)$$

If $T = |T|e^{i\delta_T}$, $R = |R|e^{i\delta_R}$, this relation becomes, by eliminating $|R|$ from (3.65),

$$\cos ka = \{|T|^2 \sin 2(\delta_T - \delta_R) - \sin 2(\kappa a + \delta_R)\}/\{2|T| \sin (\delta_T - 2\delta_R - \kappa a)\}$$

or when $\delta_T = \delta_R \pm (\pi/2)$,

$$\cos ka = \sin (\kappa a + \delta_R)/|T| \quad (3.66)$$

We note that for a square potential barrier with $E \ll V_0$ (the height of the barrier), $\delta_T = \delta_R \pm (\pi/2)$ and $\delta_R = \pm \pi/2$ (see §3.3); there will

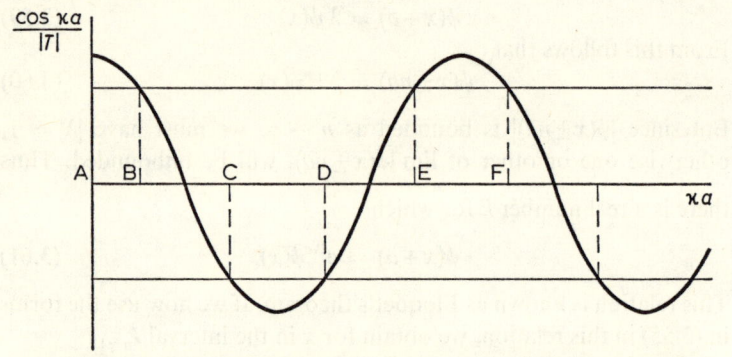

Fig. 3.12 The graph of $(\cos \kappa a)/|T|$ against κa, showing the regions for which there is no solution, called the forbidden energy regions, such as AB, CD, EF, etc.

be more general potentials for which this is true. In this case we see that since $0 \leqslant |T| \leqslant 1$, if we have $|T| < 1$ then there will be a range of values of κ for which (3.66) has no solution, i.e., those values for which $(\cos \kappa a)/|T| > 1$; these values are shown in Fig. 3.12. Such

MOTION IN ONE DIMENSION

values of κ correspond to forbidden values of the energy and are near the values $\kappa a = n\pi$. There are then *forbidden* energy regions, which can be shown to coincide with the Bragg condition for complete reflection of a beam of wave length $2\pi/\kappa$ at perpendicular incidence as follows: The general form of the Bragg condition is

$$n\lambda = 2a \sin \theta$$

where θ is the Bragg angle as given by (2.1). For perpendicular incidence, with

$$\lambda = 2\pi/k$$

we have

$$n\lambda = 2a, \quad n\pi = \kappa a$$

which are the approximate forbidden values of energy, as we have

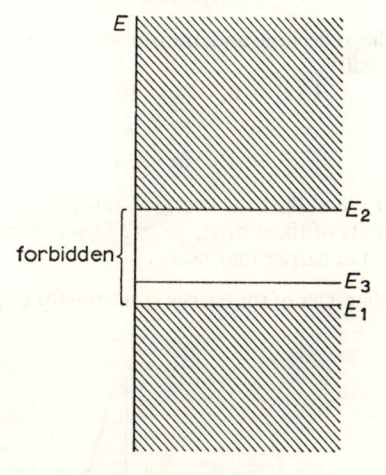

Fig. 3.13 The allowed and forbidden energy regions in a crystal. The forbidden energy region is between E_1 and E_2; E_3 is an allowed energy level arising because of impurities in the material.

seen. As we expect, the beam is excluded from the crystal when this condition is satisfied.

The typical energy levels for an electron in a lattice are shown in Fig. 3.13. If all the electron levels are filled up to energy E_1, and none filled above energy E_2, then the application of a small external electric field will not allow electrons to gain energy; so there will be no conduction of electricity, and the material will be an insulator. Any energy level in the gap between E_1 and E_2, say at E_3, will more easily allow transitions to it due to a few electrons having enough energy at room temperature to acquire energy $(E_3 - E_1)$ (but almost none to acquire energy $(E_2 - E_1)$); such materials are called semiconductors, and are the essential components of transistors. The energy levels E_3 can be obtained by the presence of impurities in the material. More generally it is by means of the existence of forbidden energy gaps that many properties of solids have been understood recently, not only on a qualitative but even a quantitative level.

PROBLEMS

3.1 Determine the reflection and transmission coefficients for the one-dimensional potential

$$V(x) = 0 \quad (x < 0)$$
$$= V_1 \quad (0 \leq x \leq a)$$
$$= V_2 \quad (a < x)$$

where $V_2 < 0 < V_1$, and the total energy E satisfies $0 < E < V_1$.

What are the limits of these as $V_1 \to \infty$? Describe a phsyical situation where such a potential barrier may occur.

3.2 Discuss the character of the energy spectrum for each of the following potential curves:

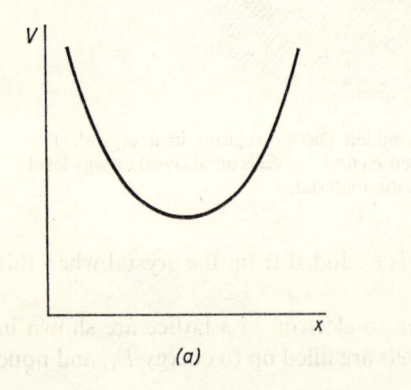

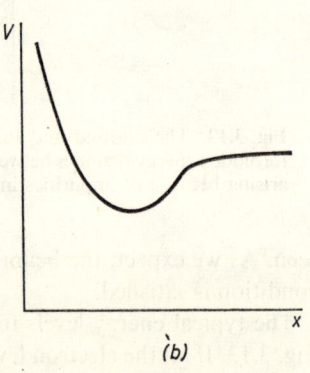

(a)

(b)

MOTION IN ONE DIMENSION 103

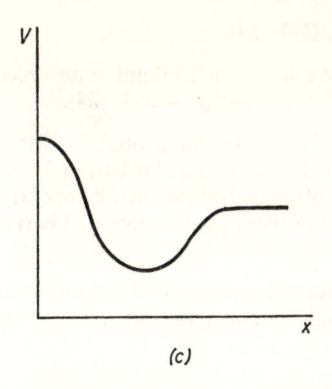

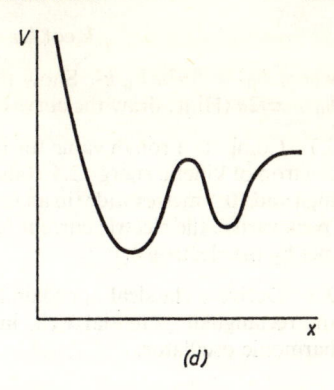

(c)

(d)

Consider also the type of wave function in the various cases (i.e., exponential or sinusoidal), and the number of free parameters entering the wave functions.

3.3 In classical mechanics the reference level for potential energy is arbitrary. What are the effects on the wave function and the energy of adding a constant potential V in Schrödinger's equation?

3.4 Compute the transmission probability of two thin rectangular barriers each of width a and height V, with

$$a \ll \hbar/(2mE)^{\frac{1}{2}}$$

separated by a distance b.

Discuss the resonance effects that can occur for certain particle energy E and separation distance b.

3.5 Show that the stationary states of a particle of mass m confined in a field free region between impenetrable walls at $x = 0$ and $x = a$ have energies E_n given by

$$E_n = \frac{n^2 h^2}{8ma^2} \quad (n = 1, 2, \cdots)$$

Obtain the corresponding wave functions.

3.6 A particle of mass m moves along the x-axis under the influence of a potential defined as

$$V(x) = \infty \quad (x < 0)$$
$$= 0 \quad (0 \leqslant x \leqslant a)$$
$$= V_0 \quad (x > a)$$

where V_0 is a positive constant. What are the continuity conditions on the wave function at $x = 0$ and $x = a$?

If the allowed energy levels E are written as $E = k^2h^2/8\pi^2m$ then there will be stationary states in which the particle is bound in the region $0 < x < a$ provided that there are real solutions of the equation

$$k \cot ka = -\sqrt{(k_0{}^2 - k^2)}$$

where $k_0{}^2 = 8\pi^2 m V_0/h^2$. Show that there is no such bound state unless $k_0 > \pi/2a$ (Hint: draw the curves $y = k \cot ka$ and $y^2 = k_0{}^2 - k^2$).

3.7 Compute a rough value for the barrier penetration probability for an electron of kinetic energy 2.5 volts incident on a rectangular barrier 3 volts high and 10^{-9} metres wide (in an ideal point contact crystal rectifier operated 'backwards', the electric current fails to flow because of a potential barrier met by the electrons).

3.8 Derive a classical approximation for the probability distribution for the rectangular potential well, in a similar manner to that used for the harmonic oscillator.

FURTHER READING

1. PEIERLS, R., *Quantum Theory of Solids*, Clarendon Press, 1955; a very clear description of the use of quantum mechanics in the theory of solids.

2. POWELL, J. and CRASEMANN, B., *Quantum Mechanics*, Addison-Wesley, 1961; especially the general potential in one dimension and Sturm-Liouville theory.

CHAPTER 4

Motion in Three Dimensions

4.1 Introduction

In the previous chapter we described the wave mechanics of a particle moving along a line. We now want to consider the more realistic case of a particle moving in space. We will do this in detail in this chapter for the special case of a potential V which has no special direction in space built into it, so that V is a function only of the radial distance r from the centre of force; the force itself will be a radial one. The lack of a special direction in space is the same as requiring that the force causing the scattering is a *spherically symmetric* one; a rotation of the co-ordinate axes with origin at the centre of force will produce the same distribution of force in space as for the unrotated situation. The Coulomb force between two charged particles is such a force field when one of the particles is regarded as the origin; the potential in this case is simply proportional to $1/r$, where r is the distance between the particles. So we may describe the motion of the electron in the hydrogen atom, the nucleus being regarded as fixed, by a spherically symmetrical potential. We will begin this chapter by considering a very important observable for a spherically symmetric system, the angular momentum. We will find out its properties, then apply this knowledge to discuss the general spherically symmetric situation, and consider the special case of the hydrogen atom. At the end of the chapter we will consider how spherical symmetry corresponds to the conservation of angular momentum, and how this relation between a symmetry and a conserved observable extends to other cases.

4.2 Angular Momentum

We defined the classical angular momentum of a particle in §1.3 by equation (1.32), $\mathbf{L} = \mathbf{r} \wedge \mathbf{p}$. We saw how to quantise a classical system in Chapter 2, and in particular we set up the rules by which $\mathbf{r}$ and $\mathbf{p}$ may be interpreted in quantum mechanics, in particular equation (2.7), $\mathbf{p} \to (\hbar/i)\nabla$. So the angular momentum observable $\mathbf{L}$ in quantum mechanics has the components

$$\left. \begin{array}{l} L_x = \dfrac{\hbar}{i}\left(y\dfrac{\partial}{\partial z} - z\dfrac{\partial}{\partial y}\right) \\[6pt] L_y = \dfrac{\hbar}{i}\left(z\dfrac{\partial}{\partial x} - x\dfrac{\partial}{\partial z}\right) \\[6pt] L_z = \dfrac{\hbar}{i}\left(x\dfrac{\partial}{\partial y} - y\dfrac{\partial}{\partial x}\right) \end{array} \right\} \quad (4.1)$$

Each of the components of $\mathbf{r}$ (or of $\mathbf{p}$) commute with the other components, $xy = yx$ etc., as they do in classical physics (though in quantum mechanics the components of $\mathbf{r}$ do *not* commute with those of $\mathbf{p}$). However, this is no longer true for $\mathbf{L}$, and we have that for the quantum mechanical angular momentum operators L_x, L_y, L_z given by (4.1), the *commutator bracket* $[L_x, L_y]_-$ between L_x and L_y or any pair of these operators is non-zero, where

$$[L_x, L_y]_- = L_x L_y - L_y L_x \quad (4.2)$$

The commutator bracket is to be evaluated as an operator acting on any wave function ψ, so that (4.2) when acting on a (suitably differentiable) wave function ψ, and using the explicit form of L_x and L_y given above, becomes

$$[L_x, L_y]_- \psi = \left(\dfrac{\hbar}{i}\right)^2 \left\{ y\dfrac{\partial}{\partial z}\left(z\dfrac{\partial \psi}{\partial x}\right) - y\dfrac{\partial}{\partial z}\left(x\dfrac{\partial \psi}{\partial z}\right) - z\dfrac{\partial}{\partial y}\left(z\dfrac{\partial \psi}{\partial x}\right) \right.$$
$$+ z\dfrac{\partial}{\partial y}\left(x\dfrac{\partial \psi}{\partial z}\right) - z\dfrac{\partial}{\partial x}\left(y\dfrac{\partial \psi}{\partial z}\right) + z\dfrac{\partial}{\partial x}\left(z\dfrac{\partial \psi}{\partial y}\right)$$
$$\left. + x\dfrac{\partial}{\partial z}\left(y\dfrac{\partial \psi}{\partial z}\right) - x\dfrac{\partial}{\partial z}\left(z\dfrac{\partial \psi}{\partial y}\right) \right\} \quad (4.3)$$

MOTION IN THREE DIMENSIONS

We now use that

$$\frac{\partial^2 \psi}{\partial x \partial z} = \frac{\partial^2 \psi}{\partial z \partial x}, \quad \frac{\partial^2 \psi}{\partial z \partial y} = \frac{\partial^2 \psi}{\partial y \partial z}, \quad \frac{\partial^2 \psi}{\partial x \partial y} = \frac{\partial^2 \psi}{\partial y \partial x},$$

so that, performing the derivatives in (4.3), we obtain for the right hand side of (4.3) the expression

$$\left(\frac{\hbar}{i}\right)^2 \left\{ yz \frac{\partial^2 \psi}{\partial z \partial x} + y \frac{\partial \psi}{\partial x} - xy \frac{\partial^2 \psi}{\partial z^2} - z^2 \frac{\partial^2 \psi}{\partial x \partial y} + xz \frac{\partial^2 \psi}{\partial y \partial z} \right.$$
$$\left. - yz \frac{\partial^2 \psi}{\partial x \partial z} + z^2 \frac{\partial^2 \psi}{\partial x \partial y} + xy \frac{\partial^2 \psi}{\partial z^2} - xy \frac{\partial^2 \psi}{\partial y \partial z} - x \frac{\partial \psi}{\partial y} \right\}$$
$$= \left(\frac{\hbar}{i}\right)^2 \left(y \frac{\partial}{\partial x} - x \frac{\partial}{\partial y} \right) \psi = i\hbar L_z \psi \qquad (4.4)$$

Combining (4.3) and (4.4) we obtain

$$[L_x, L_y]_- \psi = i\hbar L_z \psi \qquad (4.5)$$

Since the equation is valid for *all* wave functions ψ, we may rewrite the equation as an equation between operators:

$$[L_x, L_y]_- = i\hbar L_z$$

By similar evaluation we may obtain

$$[L_y, L_z]_- = i\hbar L_x$$
$$[L_z, L_x]_- = i\hbar L_y$$

These commutation relations are also often expressed in the following manner. We denote the components of **L** by (L_1, L_2, L_3), and let ε_{ijk} be the quantity

$$\begin{aligned} \varepsilon_{ijk} &= +1 \quad (i,j,k = \text{even permutation of } 1,2,3) \\ &= -1 \quad (i,j,k = \text{odd permutation of } 1,2,3) \\ &= 0 \quad (\text{otherwise}) \end{aligned} \qquad (4.6)$$

Then (4.6) is seen to be expressible as

$$[L_i, L_j]_- = i\hbar \sum_{k=1}^{3} \varepsilon_{ijk} L_k \qquad (4.7)$$

Evidently the three components of the vector **L** cannot enter into the quantum mechanical description of a particle in the same way that the position vector **r** or the momentum operator **p** do. Thus we can describe the state of a particle by its wave function $\psi(\mathbf{r})$ in which all three components of **r** enter on the same footing, and for which the values of **r** have arbitrary precision (in particular the δ-function $\delta^3(\mathbf{r}-\mathbf{r}_0)$ describes a particle *at the point* $\mathbf{r}_0$)*. However, we saw in §2.6 that since **r** and **p** do not commute we cannot have a state of the particle in which **r** and **p** are simultaneously sharp or precise; we called this result the uncertainty principle. For the same reasons, since L_x, L_y and L_z do not commute with each other, we will not be able to have a state of the particle in which L_x, L_y and L_z are simultaneously precise. This does not mean that these angular momentum operators are not of great physical importance, but it does mean that they will have to be treated more carefully than **r** or **p**. We will return to this later at the end of the chapter.

In the meantime let us consider the operator

$$\mathbf{L}^2 = L_x^2 + L_y^2 + L_z^2$$

This operator is a very important one, and we will see shortly that it describes the angular part of the Laplacian operator ∇^2 when expressed in spherical polar co-ordinates; let us first calculate the commutator brackets of $\mathbf{L}^2$ with L_x, L_y and L_z. We have

$$[\mathbf{L}^2, L_x]_- = [L_x^2, L_x]_- + [L_y^2, L_x]_- + [L_z^2, L_x]_- \qquad (4.8)$$

Since an operator commutes with any power of itself, the first term on the right hand side of (4.8) is zero; we also have that for any operators A and B:

$$[A^2, B]_- = A^2 B - BA^2$$
$$= A(AB-BA) + (AB-BA)A \qquad (4.9)$$
$$= A[A,B]_- + [A,B]_- A$$

We use (4.9) for $A = L_y, B = L_x$ and $A = L_z, B = L_x$ to obtain

$$[\mathbf{L}^2, L_x]_- = L_y[L_y, L_x]_- + [L_y, L_x]_- L_y + L_z[L_z, L_x]_- + [L_z, L_x]_- L_z$$

*$\delta^3(\mathbf{r}) = \delta(x)\delta(y)\delta(z)$ when $\mathbf{r} = (x, y, z)$.

MOTION IN THREE DIMENSIONS

We may evaluate the right hand side of this equation by means of (4.7) to give

$$[\mathbf{L}^2, L_x]_- = -i\hbar L_y L_z - i\hbar L_z L_y + i\hbar L_z L_y + i\hbar L_y L_z = 0 \quad (4.10)$$

Similarly

$$[\mathbf{L}^2, L_y]_- = [\mathbf{L}^2, L_z]_- = 0$$

$\mathbf{L}^2$ commutes with each of the components of $\mathbf{L}$. Thus we may regard $\mathbf{L}^2$ and one of the components of $\mathbf{L}$, say L_z, as simultaneously measurable, whilst L_x and L_y cannot be so regarded. Thus we can describe the state of a particle by its eigenvalues for $\mathbf{L}^2$ and L_z simultaneously; $\mathbf{L}^2$ and L_z form the *largest* set of mutually commuting operators which can be formed from L_x, L_y, L_z and their powers. This notion of the largest set of *mutually commuting* operators being used to give labels to states, the labels being the eigenvalues of the set of operators, is an extremely important one, and one which we will return to in more detail at the end of the chapter.

Let us now consider the explicit form for L_x, L_y, L_z and $\mathbf{L}^2$ in spherical polar co-ordinates so as to see the importance of these operators for discussing the Schrödinger equation. In spherical polar co-ordinates (r, θ, ϕ), as shown in Fig. 4.1,

$$\left. \begin{array}{l} x = r \sin\theta \cos\phi \\ y = r \sin\theta \sin\phi \\ z = r \cos\theta \end{array} \right\} \quad (4.11)$$

We may invert (4.11) to obtain r, θ and ϕ as functions of x, y and z:

$$\left. \begin{array}{l} r^2 = x^2 + y^2 + z^2 \\ \tan\phi = y/x \\ \cos\theta = z/(x^2 + y^2 + z^2) \end{array} \right\} \quad (4.12)$$

We may now write down $\partial/\partial x, \partial/\partial y$ and $\partial/\partial z$ in terms of $\partial/\partial r, \partial/\partial \theta, \partial/\partial \phi$ using the chain rule for differentiation:

$$\frac{\partial}{\partial x} = \frac{\partial r}{\partial x}\frac{\partial}{\partial r} + \frac{\partial \theta}{\partial x}\frac{\partial}{\partial \theta} + \frac{\partial \phi}{\partial x}\frac{\partial}{\partial \phi} \quad (4.13)$$

and similarly for $\partial/\partial y$ and $\partial/\partial z$. To calculate the right hand side of (4.13), we will need the set of partial derivatives $\partial r/\partial x, \partial r/\partial y, \partial r/\partial z$,

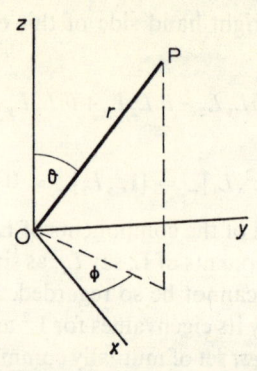

Fig. 4.1 The description of a point in three dimensions by means of the spherical polar co-ordinates r, θ, ϕ where $r = OP$, the distance of P from the origin O, θ is the angle between OP and the z-axis Oz, while ϕ is the angle between the plane perpendicular to the xy-plane through OP and the plane xOz.

$\partial\theta/\partial x$, $\partial\theta/\partial y$, $\partial\theta/\partial z$, $\partial\phi/\partial x$, $\partial\phi/\partial y$ and $\partial\phi/\partial z$; from (4.12) these are seen to be

$$\left.\begin{aligned}
\partial r/\partial x &= x/r = \sin\theta\cos\phi \\
\partial r/\partial y &= y/r = \sin\theta\sin\phi \\
\partial r/\partial z &= z/r = \cos\theta \\
\partial\theta/\partial x &= xz/(r^3\sin\theta) = (1/r)\cos\theta\cos\phi \\
\partial\theta/\partial y &= (1/r)\cos\theta\sin\phi \\
\partial\theta/\partial z &= -(1/r)\sin\theta \\
\partial\phi/\partial x &= -\sin\phi/(r\sin\theta) \\
\partial\phi/\partial y &= \cos\phi/(r\sin\theta) \\
\partial\phi/\partial z &= 0
\end{aligned}\right\} \quad (4.14)$$

Then from (4.13) and (4.14) we obtain

$$\begin{aligned}
L_z = -i\hbar r\sin\theta &\left[\cos\phi\left\{\sin\theta\sin\phi\frac{\partial}{\partial r} + \frac{1}{r}\cos\theta\sin\phi\frac{\partial}{\partial\theta} + \left(\frac{\cos\phi}{r\sin\theta}\right)\frac{\partial}{\partial\phi}\right\}\right.\\
&\left. - \sin\phi\left\{\sin\theta\cos\phi\frac{\partial}{\partial r} + \frac{1}{r}\cos\theta\cos\phi\frac{\partial}{\partial\theta} - \left(\frac{\sin\phi}{r\sin\theta}\right)\frac{\partial}{\partial\phi}\right\}\right]\\
= \frac{\hbar}{i}\frac{\partial}{\partial\phi} &
\end{aligned}$$
(4.15)

$$L_x = -i\hbar r \left\{ \sin\theta \sin\phi \left(\cos\theta \frac{\partial}{\partial r} - \frac{1}{r}\sin\theta \frac{\partial}{\partial \theta} \right) \right.$$

$$\left. - \cos\theta \left(\sin\phi \sin\theta \frac{\partial}{\partial r} + \frac{1}{r}\cos\theta \sin\phi \frac{\partial}{\partial \theta} + \frac{\cos\phi}{r\sin\theta}\frac{\partial}{\partial \phi} \right) \right\}$$

$$= i\hbar \left(\sin\phi \frac{\partial}{\partial \theta} + \cos\phi \cot\theta \frac{\partial}{\partial \phi} \right)$$

$$L_y = -i\hbar r \left\{ \cos\theta \left(\sin\theta \cos\phi \frac{\partial}{\partial r} + \frac{1}{r}\cos\theta \cos\phi \frac{\partial}{\partial \theta} - \frac{\sin\phi}{r\sin\theta}\frac{\partial}{\partial \phi} \right) \right.$$

$$\left. - \sin\theta \cos\phi \left(\cos\theta \frac{\partial}{\partial r} - \frac{1}{r}\sin\theta \frac{\partial}{\partial \theta} \right) \right\}$$

$$= i\hbar \left(\sin\phi \cot\theta \frac{\partial}{\partial \phi} - \cos\phi \frac{\partial}{\partial \theta} \right) \tag{4.17}$$

We see from (4.15) that with our choice of spherical polar coordinates, L_z is a very simple operator which we will discuss shortly; L_x and L_y are not so simple. However, if we compute the operator $L_x^2 + L_y^2 + L_z^2$ from the expressions (4.15) (4.16) and (4.17) we obtain

$$L_x^2 + L_y^2 + L_z^2$$

$$= -\hbar^2 \left\{ \frac{\partial^2}{\partial \phi^2} + \left(\sin\theta \frac{\partial}{\partial \theta} + \cot\theta \cos\phi \frac{\partial}{\partial \phi} \right) \left(\sin\phi \frac{\partial}{\partial \phi} + \cos\phi \cot\theta \frac{\partial}{\partial \phi} \right) \right.$$

$$\left. + \left(\cos\phi \frac{\partial}{\partial \theta} - \sin\phi \cot\theta \frac{\partial}{\partial \phi} \right) \left(\cos\phi \frac{\partial}{\partial \theta} - \sin\phi \cot\theta \frac{\partial}{\partial \phi} \right) \right\}$$

$$= -\hbar^2 \left\{ \frac{\partial^2}{\partial \phi^2} + (\sin^2\phi + \cos^2\phi) \frac{\partial^2}{\partial \theta^2} + (\sin^2\phi + \cos^2\phi) \cot^2\theta \frac{\partial^2}{\partial \phi^2} \right.$$

$$+ \sin\phi \cos\phi \frac{\partial(\cot\theta)}{\partial \theta} \frac{\partial}{\partial \phi} + \cos\phi \cot\theta \frac{\partial(\sin\phi)}{\partial \phi} \frac{\partial}{\partial \theta}$$

$$\left. - \sin\phi \cot\theta \frac{\partial(\cos\phi)}{\partial \phi} \frac{\partial}{\partial \theta} - \cos\phi \sin\phi \frac{\partial(\cot\theta)}{\partial \theta} \frac{\partial}{\partial \phi} \right\}$$

$$= -\hbar^2 \left(\frac{\partial^2}{\partial \theta^2} + \frac{1}{\sin^2\theta}\frac{\partial^2}{\partial \phi^2} + \cot\theta \frac{\partial}{\partial \theta} \right)$$

$$= -\hbar^2 \left\{ \frac{1}{\sin\theta}\frac{\partial}{\partial \theta}\left(\sin\theta \frac{\partial}{\partial \theta} \right) + \frac{1}{\sin^2\theta}\frac{\partial^2}{\partial \phi^2} \right\} \tag{4.18}$$

It is shown in textbooks on mathematical methods* that the form of Laplace's operator ∇^2 in spherical polar co-ordinates is

$$\nabla^2 = \frac{1}{r^2}\frac{\partial}{\partial r}\left(r^2\frac{\partial}{\partial r}\right) + \frac{1}{r^2}\left\{\frac{1}{\sin\theta}\frac{\partial}{\partial\theta}\left(\sin\theta\frac{\partial}{\partial\theta}\right) + \frac{1}{\sin^2\theta}\frac{\partial^2}{\partial\phi^2}\right\}$$

so from (4.18) we have that

$$\nabla^2 = \frac{1}{r^2}\frac{\partial}{\partial r}\left(r^2\frac{\partial}{\partial r}\right) - \frac{\mathbf{L}^2}{\hbar^2 r^2} \qquad (4.19)$$

The equation (4.19) expresses the fact that the angular part of ∇^2 is completely expressed by means of the operator $\mathbf{L}^2$, the remaining part being purely dependent on the radial distance r. We will see in the next section how this reduces the Schrödinger equation in three dimensions to a similar one in one dimension when the potential depends only on the radial distance r. Before we consider that problem we will obtain a better understanding of the angular momentum operator $\mathbf{L}$. In particular we wish to know the eigenvalues of L_x, L_y, L_z, $\mathbf{L}^2$, and also the nature of the eigenfunctions (these will be important for the hydrogen atom wave functions). We have already remarked that since L_x, L_y and L_z do not commute with each other we cannot obtain a state or wave function which is a *simultaneous* eigenfunction of these two operators. The action of the angular part of ∇^2 on such a state will then be very simple, its explicit form being given by (4.19).

Let us first consider the eigenvalues and eigenvectors of L_z. We take the wave function ψ as depending on the spherical polar co-ordinates r, θ, ϕ so that $L_z = (\hbar/i)\partial/\partial\phi$, by (4.14). Then, if ψ is an eigenfunction of L_z with eigenvalue l_z, this requires

$$(\hbar/i)\partial\psi/\partial\phi = l_z\psi \qquad (4.20)$$

The solution of (4.20) is that

$$\psi = \psi(r,\theta)e^{il_z\phi/\hbar} \qquad (4.21)$$

where $\psi(r,\theta)$ is some suitable function of r and θ. In order that there be just *one* value of the wave function at each point (r,θ,ϕ) we must have that $e^{il_z\phi/\hbar}$ is a single valued function of ϕ; if we increase ϕ by 2π the value of the wave function (4.21) must not change, so that

$$e^{il_z\phi/\hbar} = e^{il_z(\phi+2\pi)/\hbar}$$

*See, for example, "Vector Analysis" by L. Marder, no. 3. in this series.

which requires
$$l_z = m\hbar \quad (m = 0, 1, 2, \cdots) \tag{4.22}$$

Thus we see that single valuedness of the wave function implies that the z-component of angular momentum only takes integer values, in units of $\hbar$. This is a justification of the third rule used by Bohr in his model of the atom which we discussed in §2.2. It is also possible to obtain (4.22) by methods which are independent of the assumption of single valuedness of the wave function; this is contained in Appendix 1; such a method involves the use of creation and annihilation operators in a similar fashion to our discussion of the harmonic oscillator in §3.5 (where these creation and annihilation operators were $(d/dy) \mp y$).

We may also determine the eigenvalues C of $\mathbf{L}^2$ by two alternative methods, either by using the differential operator form (4.18) of $\mathbf{L}^2$ or by using creation and annihilation operator techniques. The first method involves finding the values of C for which the differential equation

$$-\hbar^2 \left\{ \frac{1}{\sin\theta} \frac{\partial}{\partial\theta} \left(\sin\theta \frac{\partial}{\partial\theta} \right) + \frac{1}{\sin^2\theta} \frac{\partial^2}{\partial\phi^2} \right\} Y(\theta,\phi) = C Y(\theta,\phi) \tag{4.23}$$

has suitably bounded solutions $Y(\theta,\phi)$ so that

$$\int_0^{2\pi} d\phi \int_0^{\pi} d\theta |Y(\theta,\phi)|^2 \sin\theta$$

is finite (leading to square integrable wave functions $R(r) Y(\theta,\phi)$ if also the radial dependence $R(r)$ is such that $\int_0^\infty |R(r)|^2 r^2 dr$ is finite). We are considering $Y(\theta,\phi)$ as a simultaneous eigenvector of L_z so that from (4.21) we see that

$$Y(\theta,\phi) = e^{im\phi} P(\theta) \tag{4.24}$$

where m is an integer. If we substitute (4.24) into (4.23) we find that $P(\theta)$ must satisfy the differential equation

$$-\hbar^2 \left\{ \frac{1}{\sin\theta} \frac{d}{d\theta} \left(\sin\theta \frac{d}{d\theta} \right) - \frac{m^2}{\sin^2\theta} \right\} P(\theta) = C P(\theta) \tag{4.25}$$

If we replace $\cos\theta$ by x, and now regard P as a function of x, with $d/dx = -(1/\sin\theta) d/d\theta$, the equation (4.25) becomes

$$\frac{d}{dx}\left\{(1-x^2)\frac{dP}{dx}\right\} + \left\{\frac{C}{\hbar^2} - \frac{m^2}{(1-x^2)}\right\}P(x) = 0 \quad (4.26)$$

This is known as Legendre's equation, and its solutions are known as Legendre functions, and denoted by $P_l^m(x)$; they are suitably bounded so that $\int_{-1}^{+1}[P_l^m(x)]^2\,dx$ is finite provided $C/\hbar^2 = l(l+1)$, $(m = 0, \pm 1, \pm 2, \cdots, \pm l)$, with l a non-negative integer. The Legendre function $P_l^m(x)$ in this case is then a polynomial in x of degree $l-|m|$, which contains only even powers of x when $l-|m|$ is even, and only odd powers of x when $l-|m|$ is odd. (We will use this property later when we discuss the reflection or parity properties of the hydrogen atom wave function.)

The alternative method of using creation or annihilation operators is more related to modern approaches in particle physics, and we will finish this section by discussing it. The method is based on the commutation rules (4.7) for the angular momentum operators. We introduce the shift (or annihilation and creation) operators $L_\pm$ by

$$L_\pm = (1/\sqrt{2})(L_x \pm iL_y)$$

Then

$$[L_+, L_-]_- = \tfrac{1}{2}[L_x + iL_y, L_x - iL_y]_-$$
$$= \tfrac{1}{2}i[L_y, L_x]_- - \tfrac{1}{2}i[L_x, L_y]_- = \hbar L_z \quad (4.27)$$

whilst

$$[L_\pm, L_z]_- = (1/\sqrt{2})[L_x, L_z]_- \pm (i/\sqrt{2})[L_y, L_z]_-$$
$$= (\hbar/\sqrt{2})[-iL_y \pm (i)^2 L_x] = \mp \hbar L_\pm \quad (4.28)$$

Let $Y(\theta, \phi)$ be a simultaneous eigenfunction of $\mathbf{L}^2$ and L_z belonging to the eigenvalues $\hbar^2 d$ and $\hbar m$ respectively; we will denote this eigenfunction as $|d, m\rangle$:

$$Y(\theta, \phi) = |d, m\rangle, \quad \mathbf{L}^2|d, m\rangle = \hbar^2 d|d, m\rangle, \quad L_z|d, m\rangle = \hbar m|d, m\rangle$$
(4.29)

We will also denote the complex conjugate wave function $Y^*(\theta, \phi)$ by $\langle d, m|$, and for two such simultaneous eigenfunctions $Y_1(\theta, \phi) = |d_1, m_1\rangle$ and $Y_2(\theta, \phi) = |d_2, m_2\rangle$ we will denote the inner product

$$\int_0^{2\pi} d\phi \int_0^{\pi} \sin\theta \, d\theta \; Y_1^*(\theta,\phi) Y_2(\theta,\phi)$$

by $\langle d_1, m_1 | d_2, m_2 \rangle$. This notation of dropping the dependence on the space variables θ, ϕ, only labelling by means of the eigenvalues of $\mathbf{L}^2$ and L_z, is due to Dirac; the states $|d, m\rangle$ are called ket vectors, the states $\langle d, m|$ bra vectors, so that the inner product $\langle d_1, m_1 | d_2, m_2 \rangle$ is naturally a bra-ket (bracket).

We will consider how the shift operators $L_\pm$ may be used to take an eigenstate of $\mathbf{L}^2$, L_z to one with different eigenvalues. Since $[L_\pm, \mathbf{L}^2]_- = 0$ (from (4.10)) then

$$\mathbf{L}^2 L_\pm |d,m\rangle = L_\pm \mathbf{L}^2 |d,m\rangle = \hbar^2 \, d L_\pm |d,m\rangle \tag{4.30}$$

Also, using (4.28), we have that

$$L_z L_\pm |d,m\rangle = [(L_z L_\pm - L_\pm L_z) + L_\pm L_z]|d,m\rangle$$
$$= (\pm \hbar L_\pm + \hbar m L_\pm)|d,m\rangle$$
$$= \hbar(m \pm 1) L_\pm |d,m\rangle \tag{4.31}$$

Thus $L_\pm |d,m\rangle$ are also simultaneous eigenstates of $\mathbf{L}^2$ and L_z with eigenvalues $\hbar^2 d$ and $\hbar(m \pm 1)$; the action of $L_\pm$ on $|d,m\rangle$ is to shift the value of m by ± 1, which is the reason that the operators $L_\pm$ are called shift operators (or creation and annihilation operators, since L_+ creates and L_- destroys a unit of angular momentum).

Let us now determine which values of d and m are allowed. We may write

$$\mathbf{L}^2 = L_z^2 + L_+ L_- + L_- L_+$$
$$= L_z^2 + 2L_+ L_- - \hbar L_z$$
$$= L_z^2 + 2L_- L_+ + \hbar L_z \tag{4.32}$$

We have seen that if we continue applying L_+ repeatedly to the wave function $|d,m\rangle$ we obtain greater and greater eigenvalues of L_z, so that

$$L_z(L_+)^r |d,m\rangle = \hbar(m+r)(L_+)^r |d,m\rangle \tag{4.33}$$

Similarly L_- applied repeatedly reduces the eigenvalues of L_z:

$$L_z(L_-)^s |d,m\rangle = \hbar(m-s)(L_-)^s |d,m\rangle \tag{4.34}$$

Can there occur arbitrarily large positive or negative eigenvalues of L_z with the same eigenvalue $\hbar^2 d$ of $\mathbf{L}^2$? We will now prove that this cannot occur. To do that let us consider the inner product

$$\langle d,m|L_+L_-|d,m\rangle$$

This is actually the inner product of $L_-|d,m\rangle$ with itself, since the wave function $(L_x-iL_y)Y(\theta,\phi)$ has $(L_x+iL_y)Y^*(\theta,\phi)$ as its complex conjugate, which we interpret as $\langle d,m|L_+$, with the differential operators in L_+ acting to the *left* on the bra $\langle d,m|$. Thus

$$\langle d,m|L_+L_-|d,m\rangle \geq 0, \quad \text{and} = 0 \text{ only if } L_-|dm\rangle = 0 \quad (4.35)$$

By similar argument

$$\langle d,m|L_-L_+|d,m\rangle \geq 0, \quad \text{and} = 0 \text{ only if } L_+|dm\rangle = 0 \quad (4.36)$$

We may rewrite (4.35) by means of (4.32) as

$$\langle d,m|(\mathbf{L}^2-L_z^2+\hbar L_z)|d,m\rangle \geq 0$$

or

$$\hbar^2(d-m^2+m)\langle d,m|d,m\rangle \geq 0$$

Since the inner product $\langle d,m|d,m\rangle$ of a state with itself is always positive, then we require

$$d-m^2+m \geq 0$$

or

$$(m-\tfrac{1}{2})^2 \leq d+\tfrac{1}{4}$$

so that

$$-\sqrt{(d+\tfrac{1}{4})}+\tfrac{1}{2} \leq m \leq \sqrt{(d+\tfrac{1}{4})}+\tfrac{1}{2} \quad (4.37)$$

Thus the range of possible values of m for a given d is bounded both above and below. This requires that for suitably large r, say a, in (4.33),

$$(L_+)^{a+1}|d,m\rangle = 0 \quad (4.38)$$

(otherwise $(L_+)^r|d,m\rangle$ is an eigenvector with eigenvalue $\hbar(m+r)$ of L_z for *all* r). Similarly there is an integer b so that in (4.34)

$$(L_-)^{b+1}|d,m\rangle = 0 \quad (4.39)$$

The range of possible eigenvalues of L_z obtained by applying the shift operators $L_\pm$ to $|d,m\rangle$ will thus be $\hbar(m-b), \hbar(m-b+1), \cdots, \hbar m, \cdots, \hbar(m+a-1), \hbar(m+a)$.

Let us now consider $\mathbf{L}^2$ applied to $|d, m+a\rangle$, and use that

$$L_+|d, m+a\rangle \propto (L_+)^{a+1}|d, m\rangle = 0 \tag{4.40}$$

from (4.38), or from (4.39) that

$$L_-|d, m-b\rangle \propto (L_-)^{b+1}|d, m\rangle = 0 \tag{4.41}$$

(the constant of proportionality is determined by requiring normalized states derived from the shifted states $(L_\pm)^r|d, m\rangle$). Then by (4.32) and (4.40)

$$\begin{aligned}\mathbf{L}^2|d, m+a\rangle &= (L_z^2 + 2L_-L_+ + \hbar L_z)|d, m+a\rangle \\ &= \hbar^2(m+a)(m+a+1)|d, m+a\rangle \\ &= \hbar^2 d|d, m+a\rangle\end{aligned}$$

so

$$d = (m+a)(m+a+1) \tag{4.42}$$

Similarly from (4.32) and (4.41)

$$\begin{aligned}\mathbf{L}^2|d, m-b\rangle &= (L_z^2 + 2L_+L_- - \hbar L_z)|d, m-b\rangle \\ &= \hbar^2(m-b)(m-b-1)|d, m-b\rangle \\ &= \hbar^2 d|d, m-b\rangle\end{aligned}$$

so that

$$d = (m-b)(m-b-1) \tag{4.43}$$

Equating the right hand sides of (4.42) and (4.43) we have

$$(m+a)(m+a+1) = (m-b)(m-b-1)$$

so that

$$m = \tfrac{1}{2}(b-a)$$

and

$$d = \tfrac{1}{2}(a+b)\{\tfrac{1}{2}(a+b)+1\}$$

We let $\tfrac{1}{2}(a+b) = l$, so that $d = l(l+1)$, and the set of values $m-b, m-b+1, \cdots, m+a$ is just the set of $(2l+1)$ values $-l, -l+1, \cdots, 0, \cdots l-1, l$, where we know that $2l = a+b$ is an integer. If we now use the discussion earlier in this section showing that the values of m are integers (or use the proof in the appendix) then l is an integer and the possible eigenvalues of $\mathbf{L}^2$ and L_z are $\hbar^2 l(l+1)$ and $\hbar m$, where m is an integer from one of the $(2l+1)$ values $-l, -l+1, \cdots, 0, \cdots, l-1, l$. This result agrees with that obtained by means of the use of the differential operator form of $\mathbf{L}^2$; the eigenfunctions $|l(l+1), m\rangle$ will thus be the *spherical harmonics*

$$|l(l+1), m\rangle = Y_l^m(\theta, \phi) = e^{im\phi} P_l^m(\cos\theta) \qquad (4.44)$$

4.3 Spherically Symmetrical Potentials

We are now ready to discuss the solutions of the Schrödinger equation for the motion in a potential which depends only on the radial distance of the particle from the centre of force. This equation will be

$$\nabla^2 \psi + \frac{2m}{\hbar^2} \{E - V(r)\} \psi(\mathbf{r}) = 0$$

or, in spherical polar co-ordinates (see (4.19))

$$\frac{1}{r^2} \frac{\partial}{\partial r}\left(r^2 \frac{\partial \psi}{\partial r}\right) - \frac{\mathbf{L}^2}{\hbar^2 r^2} \psi + \frac{2m}{\hbar^2} \{E - V(\mathbf{r})\} \psi = 0 \qquad (4.45)$$

Let us attempt to solve (4.45) by a separable solution

$$\psi(\mathbf{r}) = R(r) Y(\theta, \phi)$$

where R is a function only of the radial distance $r = |\mathbf{r}|$ and Y depends only on the angular co-ordinates θ, ϕ. The Schrödinger equation (4.45) then becomes

$$\left\{\frac{1}{r^2}\frac{d}{dr}\left(r^2\frac{dR}{dr}\right) + \frac{2m}{\hbar^2}[E - V(r)]R\right\}Y - \frac{R}{\hbar^2 r^2}(\mathbf{L}^2 Y) = 0 \qquad (4.46)$$

If we divide (4.46) throughout by RY and multiply by r^2, we obtain

$$\frac{1}{R}\frac{d}{dr}\left(r^2\frac{dR}{dr}\right) + \frac{2mr^2}{\hbar^2}[E - V(r)] = \frac{1}{\hbar^2 Y}\mathbf{L}^2 Y \qquad (4.47)$$

Since the left hand side of (4.47) is a function of r only whilst the right hand side is a function of θ, ϕ only, each side must be equal to a constant, say d. Thus

$$\mathbf{L}^2 Y = \hbar^2 d Y$$

so that $\hbar^2 d$ is an eigenvalue of $\mathbf{L}^2$. But we saw in the last section that the only eigenvalues of $\mathbf{L}^2$ for square-integrable wave functions Y are $\hbar^2 l(l+1)$ for non-negative integer l, so that $d = l(l+1)$ and

$Y = Y_l^m(\theta, \phi)$, the spherical harmonic (4.44), with m taking one of the integer values $-l, -l+1, \cdots, l-1, l$. The l is called the orbital angular momentum, m the z-component of angular momentum. The equation for the r-dependent function R is now from (4.47):

$$\frac{1}{r^2}\frac{d}{dr}\left(r^2\frac{dR}{dr}\right) + \frac{2m}{\hbar^2}\{E - V(r)\}R - \frac{Rl(l+1)}{r^2} = 0 \quad (4.48)$$

If we introduce the radial function $u = rR$, then since

$$\frac{d}{dr}\left(r^2\frac{dR}{dr}\right) = \frac{d}{dr}\left(r\frac{du}{dr} - u\right) = r\frac{d^2u}{dr^2}$$

(4.48) becomes:

$$\frac{d^2u}{dr^2} + \left[\frac{2m}{\hbar^2}\{E - V(r)\} - \frac{l(l+1)}{r^2}\right]u = 0 \quad (4.49)$$

Evidently this is in the form of a Schrödinger equation for motion in the radial variable r with potential V':

$$V'(r) = V(r) + \frac{\hbar^2}{2m}\frac{l(l+1)}{r^2} \quad (4.50)$$

This effective potential V' has a very important dependence on l; for $l = 0$, $V' = V$, whilst for $l \neq 0$ the effective potential has the term $l(l+1)/r^2$ which becomes very large and positive for small r, i.e. very repulsive. This repulsive term may combine with an attractive potential $V(r)$ to give a potential barrier away from $r = 0$; the case of an attractive step function potential for $V(r)$ is shown in Fig. 4.2. This barrier is naturally called the angular momentum barrier; it allows states to be trapped in the well so formed; these states have a finite probability of penetrating out of the angular momentum barrier, so have a finite lifetime; they are called resonance states.

Having reduced the three dimensional Schrödinger equation (4.45) to the one dimensional radial equation with effective potential V' of (4.50), we may now apply the methods we developed in Chapter 3. In general we will not have an effective potential which is one of the exactly soluble forms considered in that chapter, so that the general discussion of §3.6 is the most relevant immediately. From that discussion we may conclude that provided that E takes one of a discrete set of values $E_1, E_2, \cdots$ there will be a square-integrable solution u of this radial Schrödinger equation; we denote the eigenvalues and eigenvectors of E as $E_{n,l}$ ($n = 1, 2, \cdots$) and $u_{n,l}$, where the dependence on l is made explicit. Thus the complete wave function will be

$$\psi_{nlm}(\mathbf{r}) = \frac{1}{r} u_{n,l}(r) \, Y_l^m(\theta, \phi) \tag{4.51}$$

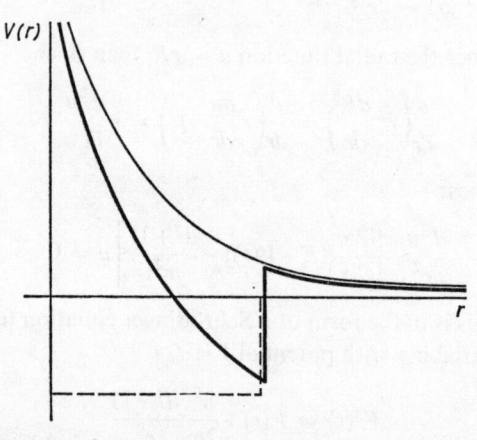

Fig. 4.2 The effective potential $V'(r) = V(r) + [l(l + 1)/r^2]$ for an attractive square well potential $V(r)$ showing the angular momentum barrier; the square well potential is shown by the broken line; the angular momentum contribution $l(l + 1)/r^2$ to V' is shown by the thin curve; the effective potential is shown by the thick curve.

We note that the energy $E_{n,l}$ does not depend on m, which takes one of the $(2l+1)$ values $-l, -l+1, \cdots, l-1, l$. Hence there are $(2l+1)$ *different* wave functions with the same energies; these wave functions are said to be *degenerate*, and the degeneracy to be $(2l+1)$-fold. The degeneracy arises due to the rotational invariance of the Hamiltonian $\{\mathbf{p}^2/2m + V(r)\}$, since then no unique direction is observable in space. But dependence of the energy on the value of L_z would generate dependence on some particular direction, so that the energy cannot depend on m. Each of the $(2l+1)$ levels with given energy $E_{n,l}$ is described by a normalized wave function (4.51) if

$$\int |\psi_{nlm}(\mathbf{r})|^2 \, d^3\mathbf{r} = \int_0^\infty |u_{nl}(r)|^2 \, dr \int_0^\pi \sin\theta \, d\theta \int_0^{2\pi} d\phi \, |Y_l^m(\theta, \phi)|^2$$
$$= 1$$

How do E_{nl} and $\psi_{nlm}(\mathbf{r})$ depend in general on n, l, m? It is usual to take E_{nl} as an increasing function of n, so that $E_{1l} < E_{2l} < \cdots$. From the discussion in §3.6, we expect that u_{nl} will have exactly $(n+1)$ zeros or nodes, one of these nodes being at infinite radial distance, the other being at $r = 0$. The node at $r = 0$ arises from the fact that if u_{nl} did not vanish at the origin, then the total wave function (4.51) would behave as $1/r$ as $r \to 0$; this can never give a solution of the Schrödinger equation (4.45) since $\nabla^2(1/r)$ is the very singular function $-4\pi\delta^3(\mathbf{r})$. As we increase l, the effective potential V' in (4.50) becomes more repulsive so that we expect the energy E_{nl} to increase due to the greater difficulty of binding. The form of the angular dependence of the waves is given exactly by the spherical harmonic $Y_l^m(\theta, \phi)$. The different values $l = 0, 1, 2, 3, 4, \cdots$ are termed $s, p, d, f, g, \cdots$ states; the shape of the s and p states are shown in Fig. 4.3, for which the spherical harmonics are

$$Y_0^0(\theta, \phi) = 0 \qquad Y_1^{-1}(\theta, \phi) = e^{-i\phi}\sin\theta$$
$$Y_1^0(\theta, \phi) = \cos\theta \qquad Y_1^1(\theta, \phi) = e^{i\phi}\sin\theta$$

where in general

$$Y_l^m(\theta, \phi) = e^{im\phi} P_l^m(\cos\theta)$$
$$P_l^m(\cos\theta) = \sin^{|m|}(\theta)\, (d^{|m|}/dx^{|m|})\, P_l(x)$$

and

$$P_l(x) = (2^l l!)^{-1}\, (d^l/dx^l)\, (x^2 - 1)^l \qquad (4.52)$$

where $x = \cos\theta$; equation (4.52) is known as Rodrigues formula.

4.4 The Hydrogen Atom

The hydrogen atom consists of an electron moving round a proton under the influence of the Coulomb attraction between the two particles. So far we have discussed the motion of only a single particle in a given potential; it is still possible to consider the hydrogen atom as a single particle system if we neglect the motion of the proton, taking it to be infinitely heavy. Such an approximation should be quite good since the proton actually is 1,836 times heavier than the electron. (The small corrections arising from the finite mass of the proton are discussed in the next chapter, which extends quantum mechanics to the motion of systems containing two or more particles.)

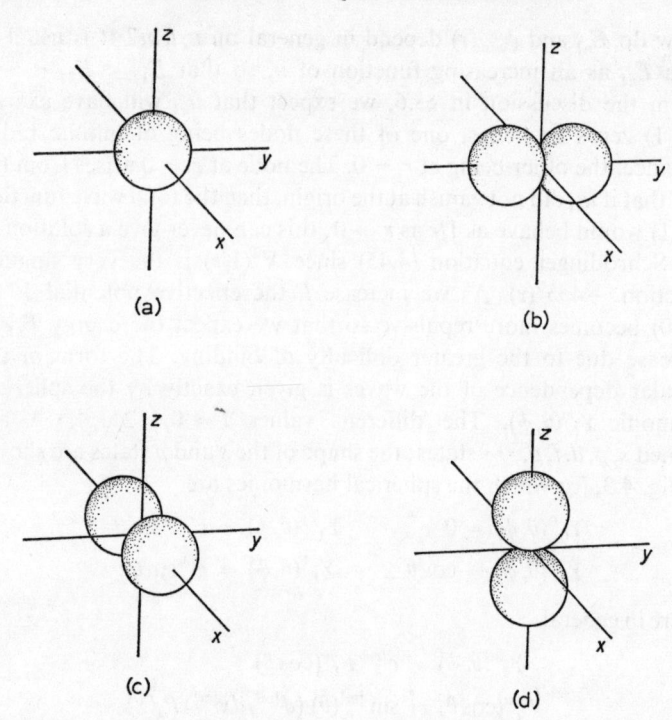

Fig. 4.3 Surfaces of constant absolute value for the s-and p-wave orbital parts of a single electron wave function. The p_x, p_y and p_z orbitals are proportional to $Y_1^{-1} + Y_1^1$, $\dfrac{1}{2i}(Y_1^1 - Y_1^{-1})$ and Y_1^0, so to x, y, and z respectively; they are cases (c), (b) and (d) respectively.

Our problem, then, is to consider the motion of the electron of mass m_e in the Coulomb field of a proton fixed at the origin. The potential function describing this motion will thus be

$$V(\mathbf{r}) = -e^2/4\pi\epsilon_0 r \tag{4.53}$$

and the Schrödinger equation is

$$\nabla^2\psi + \frac{2m_e}{\hbar^2}\left(E + \frac{e^2}{4\pi\epsilon_0 r}\right)\psi = 0 \tag{4.54}$$

This equation has a spherically symmetrical potential, so we may discuss it by the method of the previous section. In particular the solutions of (4.54) will be labelled by integers n, l, m where l and m

are the orbital angular momentum and its z-component, and by (4.50) and (4.51):

$$\psi_{nlm} = (1/r) u_{nl}(r) Y_l^m(\theta, \phi)$$

where the radial function $u_{nl}(r)$ satisfies the one dimensional Schrödinger equation

$$\frac{d^2 u_{nl}}{dr^2} + \frac{2m_e}{\hbar^2}\left[E_{nl} + \frac{e^2}{4\pi\epsilon_0 r} - \frac{\hbar^2 l(l+1)}{2m_e r^2}\right] u_{nl}(r) = 0 \quad (4.55)$$

We show the form of the effective potential entering this Schrödinger equation in Fig. 4.4. In order to find the values of E_{nl} for which $u_{nl}(r)$

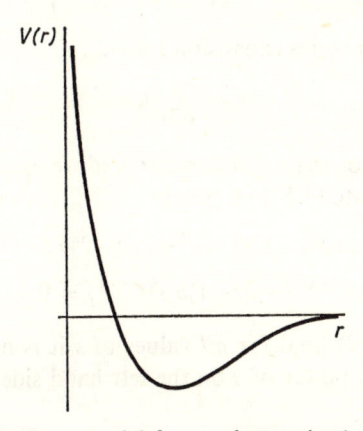

Fig. 4.4 The effective potential for an electron in the hydrogen atom.

is square integrable, we may again use either the differential equation directly or, alternatively, creation and annihilation operator techniques. Since we already have the differential equation it is simpler to use it directly. Let us first reduce the number of constants floating around in the equation. We define a new variable s and a new 'energy' λ by

$$E = \frac{\lambda m_e e^4}{2\hbar^2 (4\pi\epsilon_0)^2}, \quad r = \frac{4\pi\hbar^2 \epsilon_0 s}{m_e e^2}$$

so that (4.55) becomes (dropping the labels n, l on u and E)

$$\frac{d^2 u}{ds^2} + \left[\lambda + \frac{2}{s} - \frac{l(l+1)}{s^2}\right] u(s) = 0 \quad (4.56)$$

where u is now regarded as a function of the new variable s. Now for large s (4.56) reduces to $(d^2u/ds^2)+\lambda u = 0$, so that u behaves like $e^{-s\sqrt{-\lambda}}$ as s becomes large (where we expect the bound state energy E to be negative, so that λ will also be negative). Let us remove this exponential behaviour by considering a new function ω with $u = e^{-s\sqrt{-\lambda}}\omega$. Then ω must satisfy the equation

$$\frac{d^2}{ds^2}(e^{-s\sqrt{-\lambda}}\omega)+e^{-s\sqrt{-\lambda}}\left\{\lambda+\frac{2}{s}-\frac{l(l+1)}{s^2}\right\}\omega = 0$$

or

$$\frac{d^2\omega}{ds^2}-2\sqrt{-\lambda}\frac{d\omega}{ds}+\left\{\frac{2}{s}-\frac{l(l+1)}{s^2}\right\}\omega = 0 \qquad (4.57)$$

Let us try a power series expansion for ω:

$$\omega = \sum_{n\geqslant 0} a_n s^{n+\alpha} \qquad (4.58)$$

where α is some non-zero real number, and we assume $a_0 \neq 0$. If we substitute (4.58) into (4.57) we obtain

$$\sum_{n\geqslant 0}\{a_n(n+\alpha)(n+\alpha-1)s^{n+\alpha-2}-2\sqrt{(-\lambda)}a_n(n+\alpha)s^{n+\alpha-1}$$
$$+2a_n s^{n+\alpha-1}-l(l+1)a_n s^{n+\alpha-1}\} = 0 \qquad (4.59)$$

In order that (4.59) hold for *all* values of s it is necessary that the coefficient of each power of s on the left hand side of (4.59) vanish. This gives:

(i) for the coefficient of $s^{\alpha-2}$:

$$\alpha(\alpha-1) = l(l+1)$$

so that

$$\alpha = -l \quad \text{or} \quad \alpha = l+1$$

Since we know that u must vanish at $s = 0$ (see the argument at the end of the previous section) then we must choose $\alpha = l+1$.

(ii) for the coefficient of $s^{n+\alpha-2}$:

$$a_n\{(n+\alpha)(n+\alpha-1)-l(l+1)\} = 2a_{n-1}\{(n+\alpha-1)\sqrt{(-\lambda)}-1\}$$

If we substitute $\alpha = l+1$ in this equation we obtain

$$a_n = 2a_{n-1}\frac{\{(n+l)\sqrt{(-\lambda)}-1\}}{(n+l+1)(n+l)-l(l+1)} \qquad (4.60)$$

Then for large n, a_n/a_{n-1} behaves like $2\sqrt{(-\lambda)}/n$, which is the same ratio of coefficients of succeeding powers of s as arise in the expansion of $e^{2s\sqrt{-\lambda}}$. But then $u(s)$ would behave as $e^{s\sqrt{-\lambda}}$ for large s, which is not a square integrable function. To prevent this we have to require that a_n vanishes for some n (so far all succeeding n, by (4.60)). There must thus be an integer n_r, so that

$$(l+n_r)\sqrt{-\lambda} = 1, \quad a_{n_r} = a_{n_r+1} = \cdots = 0$$

so

$$E_{n_r,l} = \frac{m_e e^4}{2\hbar^2(4\pi\epsilon_0)^2(l+n_r)^2} \tag{4.61}$$

It is usual to introduce the *principle quantum number* n instead of n_r, with $n = n_r + l$, so that $E_{n_r,l}$ depends only on n, whilst l can take the values $0, 1, 2, \cdots, (n-1)$ (since n_r is a *strictly* positive integer). We see that there is a degeneracy, since $E_{n_r,l}$ only depends on n (agreeing exactly with the Bohr values of §2.2) and is independent of l; we denote the energy as E_n. Thus for each value of n there will be n values of l for which the wave functions ψ_{nlm} have energy E_n; each of these functions may have any one of $(2l+1)$ values of m, for each l, again not affecting the energy of the corresponding motion of the electron. There are thus $\sum_{l=0}^{(n-1)}(2l+1) = n^2$ states with a given principal quantum number n and energy E_n, but different wave-functions. This extra

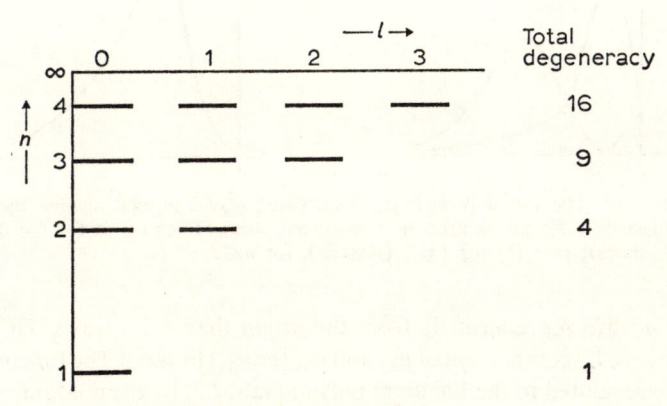

Fig. 4.5 The energy levels E_n and their degeneracy n^2 for electrons in the hydrogen atom; the explicit levels shown are for $n = 1, 2, 3$ and 4.

degeneracy beyond that arising for spherically symmetrical potentials which we found in the previous section is due to the $(1/r)$ form of the Coulomb potential; the answer to Problem 4.2 shows that this degeneracy is removed for other spherically symmetrical potentials. The levels and their degeneracy are shown in Fig. 4.5.

The function $\omega_{n,l}(s)$, regarded now as depending on n and l, will be a polynomial of degree $(n_r-1)+(l+1) = n$ (remembering the factor s^{l+1} multiplying each term $a_n s^n$ in (4.58)). Then $\omega_{n,l}$ will have n nodes, and $u_{n,l}$ will have $(n+1)$ nodes, one being at infinity. We quote the functions $u_{n,l}$ for $n = 1$ and 2 (to within normalization constants):

$$\left.\begin{aligned} u_{10} &= se^{-s} \\ u_{20} &= se^{-s/2}(2-s), \qquad u_{21} = s^2 e^{-s/2} \\ u_{30} &= s(27-18s+2s^2)e^{-s/3}, \qquad u_{31} = s^2(6-s)e^{-s/3}, \\ u_{32} &= s^3 e^{-s/3} \end{aligned}\right\} \quad (4.62)$$

and give their radial probability distributions (correctly normalized) in Fig. 4.6. We see that the p-wave ($l = 1$) is more highly excluded from the origin than the s-wave, as is to be expected from the angular momentum barrier; this barrier gets stronger as l increases so repels

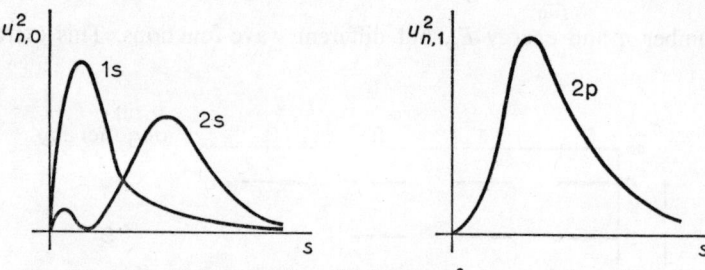

Fig. 4.6 The radial probability distribution $u_{nl}^2(s)$ plotted against the distance s for an electron in a hydrogen atom. Case (a) is for $l = 0$ (s-states); case (b) for $l = 1$ (p-states), for $n \leqslant 2$.

the p-wave more strongly from the origin than the s-wave. This is also seen by comparison of u_{32} and u_{31} (or u_{21}) in (4.62). The functions $u_{n,l}$ are related to the Laguerre polynomials, L_r^s, by the relation

$$u_{n,l}(s) = e^{-s/n} s^{l+1} L_{n-l-1}^{2l+1}(2s/n) \quad (4.63)$$

4.5 Time Dependence *

In the first chapter we saw that there are certain quantities which were constant or independent of the time for classical systems; such quantities were the total energy or momentum, and the orbital angular momentum for a particle moving in a central force. We want to discuss briefly how the idea of conserved quantities or constants of the motion arises in quantum mechanics, and to see how such quantities may play a useful role in describing general aspects of the motion in the same fashion that constants of the motion could be used in classical mechanics. In order to do this, we will consider, in the present section, how the development with time of observables is described in quantum mechanics. This will be used in the next section to discuss constants of the motion, for the special case of motion in three dimensions.

We have used the time dependent Schrödinger equation

$$i\hbar \frac{\partial}{\partial t} \psi(\mathbf{r}, t) = H\psi(\mathbf{r}, t) \qquad (4.64)$$

many times so far, but always reduced the time dependence of $\psi(\mathbf{r}, t)$ to the most trivial one, that for stationary states, with $\psi(\mathbf{r}, t) = e^{-iEt/\hbar}\psi(\mathbf{r}, 0)$, where E is an eigenvalue of the Hamiltonian H. But not all states are stationary states, furthermore, in many situations when the state is nearly stationary, we may be interested in the probability of the transition to another stationary state with the emission or absorption of energy. In order to proceed with such problems we will consider here the general problem of the time dependence of the wave function $\psi(\mathbf{r}, t)$ which satisfies the time dependent Schrödinger equation (4.64).

Let us first recognise that the description of a physical system, such as a single particle moving in a field of force, by means of (4.64) is very lopsided. For we ascribe *all* the time dependence to the wave function $\psi(\mathbf{r}, t)$ whilst the observables, in particular the position $\mathbf{r}$ and momentum $(\hbar/i)\nabla$ are independent of time. This mode of description of the time development is called the *Schrödinger picture*. There is an alternative mode of description, which is equally lopsided but to the other extreme: the wave function is independent of time, whilst the dynamical variables contain all the time dependence; this picture is called *the Heisenberg picture*. To obtain the Heisenberg from the Schrödinger picture, we have to transform our wave functions so as to remove their time dependence. If we denote by $\psi_s(t)$ the wave function which satisfies the time dependent Schrödinger equation (4.64), then we may solve (4.64) formally as

$$\psi_s(t) = e^{-iHt/\hbar}\psi_s(0) \qquad (4.65)$$

The exponential $e^{-iHt/\hbar}$ is defined by a power series expansion, and the operator acts on wave functions for which this power series converges. The wave function $\psi_s(0)$ is the Schrödinger wave function at time $t = 0$; differentiating (4.65) formally gives

$$i\hbar \frac{\partial}{\partial t}\psi_s(t) = i\hbar(-iH/\hbar)\, e^{-iHt/\hbar}\psi_s(0) = H\psi(t)$$

as required. We see then that the wave function $e^{iHt/\hbar}\psi_s(t)$, being equal to $\psi_s(0)$, is independent of time; we may take this transformed wave function to be the Heisenberg picture wave function ψ_H:

$$\psi_H = e^{iHt/\hbar}\psi_s(t) \qquad (4.66)$$

We do not specify the time dependence of ψ_H, since as we have seen ψ_H is time independent. We see then that the operator $e^{iHt/\hbar}$ achieves the transformation from the Schrödinger picture to the Heisenberg picture, at least for wave functions. What about the corresponding transformation for the dynamical variables? We wish to transform them so that the *value* of a dynamical variable does not depend on the picture in which it appears. If a dynamical variable is represented by an operator A_s in the Schrödinger picture, and $A_H(t)$ in the Heisenberg picture (where we make explicit the time dependence of the latter operator) and if the Schrödinger and Heisenberg picture wave functions $\psi_s(t)$ and ψ_H are related by (4.66), we require that the Schrödinger picture expectation value $(\psi_s(t), A_s\psi_s(t))$ and the corresponding Heisenberg picture expectation value $(\psi_H, A_H(t)\psi_H)$ are equal:

$$(\psi_s(t), A_s\psi_s(t)) = (\psi_H, A_H(t)\psi_H)$$

If we now use the relation between $\psi_s(t)$ and ψ_H expressed by (4.66), but in the form $\psi_s(t) = e^{-iHt/\hbar}\psi_H$ obtained from (4.66) by inverting the exponential operator, we have

$$(e^{-iHt/\hbar}\psi_H, A_s e^{-iHt/\hbar}\psi_H) = (\psi_H, A_H(t)\psi_H) \tag{4.67}$$

But since $(e^{-iHt/\hbar}\psi, e^{-iHt/\hbar}\psi) = (\psi, \psi)$ or $e^{-iHt/\hbar}$ is a unitary operator, not changing the 'length' $(\psi, \psi)^{\frac{1}{2}}$ of the wave function ψ on which it acts (we may regard it essentially as a phase factor) then we may rewrite the right-hand side of (4.67) as $(\psi_H, e^{iHt/\hbar} A_s e^{-iHt/\hbar}\psi_H)$, so that (4.67) becomes

$$(\psi_H, e^{iHt/\hbar} A_s e^{-iHt/\hbar}\psi_H) = (\psi_H, A_H(t)\psi_H) \tag{4.68}$$

But then we may drop the wave functions ψ_H on both sides of (4.68) since it is assumed true for *all* ψ_H, and there are very many of them, to give the operator equation

$$A_H(t) = e^{iHt/\hbar} A_s e^{-iHt/\hbar} \tag{4.69}$$

We have thus obtained the relation between the Schrödinger picture and Heisenberg picture operators which represent a given dynamical variable. From (4.69) we may deduce *Heisenberg's equation of motion* for $A_H(t)$: we differentiate both sides of (4.69) to obtain

$$i\hbar \, dA_H(t)/dt = -He^{iHt/\hbar} A_s e^{-iHt/\hbar} + e^{iHt/\hbar} A_s e^{-iHt/\hbar} H$$
$$= [A_H(t), H]_- \tag{4.70}$$

This equation of motion is to be regarded as the Heisenberg picture equivalent of the Schrödinger wave equation in the Schrödinger picture. We may use it to determine the dynamical development of the system, as was done by Heisenberg in his formulation of quantum mechanics. This formulation was called *matrix mechanics* because the operators $A_H(t)$ were regarded as infinite dimensional matrices. We may obtain such matrices by taking the 'matrix elements' of $A_H(t)$ between the eigen-states $|n\rangle$ of the Hamiltonian H, for which $H|n\rangle = E_n|n\rangle$ (all the other labels in the ket $|n\rangle$ have been dropped, similarly to our discussion of bras and kets for angular momentum operators in §4.2.). Then if we take the matrix elements of both sides of (4.70), that is we sandwich both sides between a bra $\langle n|$ and a ket $|m\rangle^{(*)}$, we obtain

*This 'sandwich' process involves integration over the variables on which the wave function depends, as we saw earlier when we denoted the angular wave

$$i\hbar\, d/dt\, \langle n|A_H(t)|m\rangle = \langle n|[A_H(t)H - HA_H(t)]|m\rangle$$
$$= (E_m - E_n)\langle n|A_H(t)|m\rangle$$

Thus we may solve this differential equation to give

$$\langle n|A_H(t)|m\rangle = e^{-i(E_n - E_m)t/\hbar}\langle n|A_H(0)|m\rangle \qquad (4.71)$$

The time dependence of this matrix element has frequency $\omega = (E_m - E_n)/\hbar$, which is exactly that given by Bohr's frequency condition; it is possible to obtain these frequencies by diagonalizing the matrix $\langle n|H|m\rangle$; the diagonal values of this diagonalized matrix will then be the energy levels E_n since these diagonal values are precisely the eigenvalues of the operator H represented by the matrix $\langle n|H|m\rangle$.

We will not discuss this matrix mechanics further here since it is not as simple or useful as the wave mechanical form; we will, however, continue using the Heisenberg picture to discuss conserved quantities.

4.6 Symmetries and Conservation Laws *

As the first conservation law for our system of a particle moving under the influence of some field, let us consider the change in time of the energy of the particle. In this case the Heisenberg and Schrödinger picture operators are identical, since

$$e^{iHt/\hbar}He^{-iHt/\hbar} = H$$

Then evidently the Hamiltonian is a constant in the Heisenberg picture, as we can also see from the Heisenberg equation of motion:

$$i\hbar\, dH/dt = [H, H]_- = 0$$

This means that any wave function which is an eigenstate of H at a

---

functions $Y(\theta, \phi)$ by the ket $|d, m\rangle$, where $\hbar^2 d$ and $\hbar m$ were the eigenvalues of L^2 and L_z taken by $Y(\theta, \phi)$; the sandwiching of an operator between bras and kets of this type still involves integration over the angular variables θ and ϕ, as for example in

$$\langle d, m|(\hbar/i)\,\partial/\partial\phi|d'm'\rangle = \int Y^*(\theta, \phi)(\hbar/i)\,\partial/\partial\phi\, Y'(\theta, \phi)\,d\Omega$$

More generally the kets $|n\rangle$ correspond to wave functions which have the same labels n as the kets; the sandwich process will involve integration over the same variables as those upon which the wave functions depend.

given time remains an eigenstate for all times; we have already used this property to justify the stability of matter by wave mechanics. Indeed we have used the eigenvalue of the energy as a *label* for a state of the system described by such a wave function; this label is correct for *all* times. Such a labelling of states by means of the energy is very important because of this time independent quality of the label: the label is the same before and after any reaction. Thus if a particle is scattered by a force field, we know that the initial and final energies of the particle are equal. In this way we may give some restrictions on the motion.

It is evidently of interest to find other observables which are constant in time, so that we may use their eigenvalues as further labels of this kind for the physical states, these labels specifying the motion even further. We have already pointed out in relation to the uncertainty principle, that we can only have simultaneous eigenvectors for two operators if these operators commute. Thus we can only have simultaneous labels for eigenstates of operators H and A (some observable) if

$$[H, A]_- = 0 \qquad (4.72)$$

But the condition that H and A commute is precisely that which implies that A is time independent, since then

$$A_H(t) = e^{iHt/\hbar} A e^{-iHt/\hbar} = A$$

and

$$i\hbar \, dA_H/dt = [A_H, H]_- = [A, H]_- = 0$$

Thus the eigenvalues of A are also good labels for physical states, being constant in time, and so helping to specify the motion. Operators A which commute with the Hamiltonian H as in (4.72) are thus constants of the motion and their eigenvalues are good labels, or, as is usually stated, *good quantum numbers*.

As an example of this idea of good quantum numbers let us consider the commuting operators $\mathbf{L}^2, L_z$. If we have that

$$[H, \mathbf{L}]_- = 0 \qquad (4.73)$$

then evidently $[H, \mathbf{L}^2]_- = [H, L_z]_- = 0$, so that $H, \mathbf{L}^2, L_z$ form a mutually commuting set of operators. Hence their simultaneous eigenvalues $E, \hbar^2 l(l+1), \hbar m$ are good quantum numbers, and may be used to label the states as $|E, l, m\rangle$. We have already obtained an explicit form of such states when the system considered is an electron moving in the Coulomb field of an infinitely heavy nucleus; they are given by the wave functions of (4.51) and (4.63):

$$|E_n, l, m\rangle = \frac{1}{r} u_{nl}(r) Y_l^m(\theta, \phi) \qquad (4.74)$$

We have not, however, shown that (4.73) is valid in this case, so we are not sure that the labels are good quantum numbers, i.e., the orbital angular momentum and its z-component are conserved. We will now show this for the case of any spherically symmetrical potential:

$$H = (\mathbf{p}^2/2m) + V(r) \qquad (4.75)$$

Let us evaluate $[H, L_z]_-$, using for L_z the operator of (4.15): $L_z = (\hbar/i)\partial/\partial\phi$. We have that

$$[H, L_z]_- = [\mathbf{p}^2, L_z]_- (1/2m) + [V(r), L_z]_- \qquad (4.76)$$

and we will evaluate the two terms on the right-hand side of (4.76) separately. Now $\mathbf{p}^2 = -\hbar^2 \nabla^2$, and we use for ∇^2 the differential operator (4.19); but then evidently $[\nabla^2, \partial/\partial\phi]_-\psi = 0$ for *any* wave function ψ, so that $[\mathbf{p}^2, L_z]_- = 0$. For the second term on the right hand side of (4.76) exactly the same argument is valid, so that

$$[H, L_z]_- = 0.$$

We could have chosen the line with respect to which the angle ϕ is measured and the plane in which θ is measured differently, say as the x-axis and the plane yz, or the y-axis and the plane xz; these choices would give $L_x = (\hbar/i)\partial/\partial\phi$ or $L_y = (\hbar/i)\partial/\partial\phi$ respectively, and by similar arguments we have that $[H, L_x]_- = [H, L_y]_- = 0$. This proves (4.73) for the Hamiltonians of the form (4.75), and so in particular proves that l and m are good quantum numbers in the hydrogen atom states (4.74).

We have been using the phrase 'spherically symmetric' for systems with Hamiltonians of the form of (4.75), for which there is no direction singled out in space. We can relate this idea of a symmetry to the conservation of angular momentum as expressed by (4.73). We do this by considering the *group* of transformations which act on the wave functions as if a rotation of the system had been performed. How is the rotation of a wave function effected? Well, first of all we mean by such a rotation that we evaluate the wave function at the rotated point. Let us consider for simplicity a rotation through a small angle θ about the z axis. Then the point (x, y) (dropping the z-coordinate), is rotated to the value $(x - \theta y, y + \theta x)$, and so our new wave function is $\psi(x - \theta y, y + \theta x)$. This can be written as

$$\begin{aligned} \psi(x - \theta y, y + \theta x) &= \psi(x, y) + \theta(x\, \partial/\partial y - y\, \partial/\partial x)\psi(x, y) \\ &= \psi(x, y) + i\theta L_z \psi(x, y)/\hbar \\ &\approx e^{i\theta L_z/\hbar} \psi(x, y) \end{aligned} \qquad (4.77)$$

It may be shown that for non-infinitesimal angles θ the exponential expression in (4.77) is correct, since now we are interested in $\psi(x\cos\theta - y\sin\theta, y\cos\theta + x\sin\theta)$ and that may be shown to be exactly equal to $e^{i\theta L_z/\hbar}\psi(x, y)$. The rotations about the z-axis form a *group*, where the product of two rotations is the rotation about the z-axis through the sum of the angles of the two separate rotations. This group is correctly *represented* by the exponential operators $e^{i\theta L_z/\hbar}$, since

$$e^{i\theta_1 L_z/\hbar} e^{i\theta_2 L_z/\hbar} = e^{i(\theta_1 + \theta_2) L_z/\hbar} \qquad (4.78)$$

Now the 'length' (or norm) of the wave function ψ and the rotated wave function $e^{i\theta L_z/\hbar}\psi$ are evidently the same, as may be seen by direct evaluation; this means that $e^{i\theta L_z/\hbar}$ is a unitary operator. Moreover this operator does not alter scalar products between wave functions:

$$(e^{i\theta L_z/\hbar}\psi, e^{i\theta L_z/\hbar}\phi) = (\psi, \phi) \qquad (4.79)$$

Then we may regard rotations as *generated* by the operator L_z, which is called the generator of the group of rotations about the z-axis. In a similar fashion the operator of rotations through an angle θ about an axis **n** in space is achieved for wave functions by $e^{i\theta \mathbf{n}\cdot \mathbf{L}/\hbar}$, so that these rotations are generated by the operators **L**, which are thus the generators of the rotation group in space. If we compare the development with time of the original wave function and the rotated wave function, we see that they need not by any means be described by the same Hamiltonian. Thus if $\psi = e^{i\theta L_z/\hbar}\phi$, then

$$i\hbar\, \partial\psi/\partial t = e^{i\theta L_z/\hbar} i\hbar\, \partial\phi/\partial t = e^{i\theta L_z/\hbar} H\phi = e^{i\theta L_z/\hbar} H e^{-i\theta L_z/\hbar} \psi \qquad (4.80)$$

Only if H commutes with L_z will the rotated wave function ψ have the same Hamiltonian H describing its time development as the unrotated wave function ϕ. If that is the case, and if we rotate *all* wave functions simultaneously, it will not be possible to make any physical measurement which will show that such a simultaneous rotation has been made, i.e., no direction in space will be singled out. On the other hand, if H does not commute with L_z then the rotated wave functions will have a different development with time from the unrotated wave functions, and it should be possible to see the effect of this simultaneous rotation. Generalizing to arbitrary rotations in space we see that if $[H, \mathbf{L}]_- = 0$, then an arbitrary rotation of all wave functions is not *dynamically* observable, and the system is said to possess *rotation invariance* (or be spherically symmetrical); otherwise it is said to be non-invariant under rotations. Rotation invariance is found to be satisfied down to the very smallest distances of the order of 10^{-16} metres at which elementary particle interactions have been studied; this means that the angular momentum provides very good quantum numbers to label states.

Another very good invariance is that under translations. We have already seen how time translations may be achieved by means of the operator $e^{-iHt/\hbar}$:

$$\psi_s(t) = e^{-iHt/\hbar}\psi_s(0)$$

The time translations of the wave functions form a group; the operator describing the action of these translations on the wave functions is, as we have said, $e^{-iHt/\hbar}$, so that H is the generator of the time translations. The law of conservation of energy is interpreted by the above line of argument as the statement of invariance under time translations: a time translated wave function (so at a later time) has the same Hamiltonian as the untranslated wave function (which is a property of the description of the development with time of the wave function). This time translation invariance is not satisfied if the Hamiltonian H is not only a function of time through its dependence on $\mathbf{r}$ and $\mathbf{p}$ (which depend on time) but also depends on time explicitly, so that

$$H = H(\mathbf{r}, \mathbf{p}, t)$$

Such explicit time dependence could arise from an external field influencing the motion of the particle; there is no evidence of such influence at the level of sub-nuclear particles, nor at any other level.

We may also consider space translations. These may be achieved for wave functions as follows. The translation $\mathbf{r} \to \mathbf{r} + \mathbf{a}, \psi(\mathbf{r}) \to \psi(\mathbf{r} + \mathbf{a})$

is given by
$$\psi(\mathbf{r}+\mathbf{a}) = e^{\nabla \cdot \mathbf{a}}\psi(\mathbf{r})$$
(by Taylor's expansion in **a** about **r**). Hence
$$\psi(\mathbf{r}+\mathbf{a}) = e^{i\mathbf{p} \cdot \mathbf{a}/\hbar}\psi(\mathbf{r}) \tag{4.81}$$
and we may regard **p** as the generator of space translations. These translations will form an invariance group, or the system will be translation invariant, if
$$[H, \mathbf{p}]_- = 0 \tag{4.82}$$

This is evidently so for a free particle, as is to be expected, since the particle has no point in space singled out for it. This is not the case for a particle moving near a fixed centre of force, since the centre of force is a distinguished point in space. So in the former case momentum is a good quantum number, whilse in the latter case it is not. Again, translation invariance is found to be valid down to 10^{-16} metres; there are no fixed centres of force.

We will finish this section by considering a transformation which was found in 1957 to be a non-invariance of physical systems. This is the transformation of reflection in the origin of co-ordinates: $\mathbf{r} \to -\mathbf{r}$. Evidently the Hamiltonian (4.75) is unchanged under this transformation, so that the transformation should provide good quantum numbers to label states with such Hamiltonians. The label in this case is called the *parity* of the state. If we denote by π the operator on wave functions:
$$\psi(\mathbf{r}) \to \psi(-\mathbf{r}) = \pi\psi(\mathbf{r}) \tag{4.83}$$
then
$$\pi^2 = 1$$

Also π is self-adjoint: $(\psi, \pi\phi) = (\pi\psi, \phi)$ for any wave functions ϕ and ψ. Thus π is measurable and has real eigenvalues. Since $\pi^2 = 1$ then these eigenvalues are ± 1; a state with $\pi = \pm 1$ is said to be even or odd under reflection. Let us consider the parity of the wave functions for an electron in a hydrogen atom,
$$\psi_{nlm} = \frac{1}{r}u_{nl}(r)\,e^{im\phi}P_l^m(\cos\theta)$$

Under $\mathbf{r} \to -\mathbf{r}$ we have that the spherical polar co-ordinates transform as

$$(r, \theta, \phi) \to (r, \pi - \theta, \pi + \phi)$$

as we see in Fig. 4.1. Hence

$$\pi \psi_{nlm} = \frac{1}{r} u_{nl}(r) e^{im(\pi + \phi)} P_l^m(-\cos\theta)$$

Since $P_l^m(\cos\theta)$ is a polynomial of degree $(l-|m|)$ which contains only even powers of $\cos\theta$ when $l-|m|$ is even, or odd powers of $\cos\theta$ when $l=|m|$ is odd, then

$$\pi \psi_{nlm} = (-1)^m (-1)^{l-|m|} \psi_{nlm} = (-1)^l \psi_{nlm}$$

and the parity of ψ_{nlm} is $(-1)^l$, so is even for $s, d, \cdots$ waves and odd for $p, f, \cdots$ waves. The discovery of parity violation in 1957 caused a new look to be taken at the conservation laws of energy, angular momentum, linear momentum and other quantities. The ones discussed here, other than parity, have not fallen yet; they are not, however, sacrosanct, even though they have lasted over three hundred years, being modified in the way we have seen in the translation from classical to quantum mechanics.

PROBLEMS

4.1 Solve the Schrödinger equation for the two dimensional isotropic harmonic oscillator in both rectangular and polar co-ordinates, and show that the energy levels are $E_n = (n+1)\hbar w$, each energy level being $(n+1)$-fold degenerate.
[The isotropic oscillator has potential $V(x, y) = \frac{1}{2}k(x^2 + y^2)$].

4.2 Show that the modified Coulomb potential $V = -e^2/4\pi\epsilon_0 r(1 + a/r)$, where a is a constant, removes the accidental degeneracy of the hydrogen atom, and obtain the corresponding energy levels.

4.3 Assume that a single particle has orbital angular momentum with a z-component of $m\hbar$ and a square magnitude of $l(l+1)\hbar^2$. Show that

(i) $\langle L_x \rangle = \langle L_y \rangle = 0$

(ii) $\langle L_x^2 \rangle = \langle L_y^2 \rangle = \frac{1}{2}[l(l+1) - m^2]\hbar^2$

4.4 The wave function of a particle of mass m moving in a potential well at a particular time is

$$\psi(\mathbf{r}) = (x + y + z)e^{-\alpha r}$$

Calculate the probability of obtaining, for a measurement of L^2 and L_z, the results $2\hbar^2$ and 0 respectively.

4.5 A particle is confined in a sphere of radius a by an infinitely high potential barrier at the surface of the sphere, but moves freely within the sphere. Find the wave functions and energy levels for the spherically symmetrical stationary states. Show that the particle, when in the lowest energy state, is equally likely to be found inside or outside a distance $a/2$ from the centre.

4.6 A particle of mass m is confined by an infinite potential barrier to move freely within the cube $0 < x < a, 0 < y < a, 0 < z < a$. What are the wave functions for the stationary states? Calculate the corresponding energy, and verify that the lowest energy level is non-degenerate but that the next higher level is triply degenerate.

4.7 A particle of mass m moves in a spherically symmetrical potential $V = A/r^2 - B/r$, where A and B are constants. Show that the radial part G of the wave function for a bound state with energy E and angular momentum $l\hbar$ is

$$G = \rho^2 e^{-\frac{1}{2}\rho} F(\rho)$$

where $\rho = -(8mE)^{\frac{1}{2}} r/\hbar$, $(2s+1) = [(2l+1)^2 + 8mA/\hbar^2]^{\frac{1}{2}}$ and F is a polynomial in ρ. Determine the bound state energy values.

4.8 A particle of mass m moves in an attractive potential

$$V = -V_0 \quad (r < a)$$
$$= 0 \quad (r \geq a)$$

If bound s- and p-states exist, show that their energies are

$$E = -\hbar^2 \eta^2 / 2ma^2$$

where $\xi \cot \xi = -\eta$ (for s-states) and $\xi \cot \xi = 1 + (\xi^2/\eta^2)(1+\eta)$ (for p-states) and in both cases

$$\xi^2 + \eta^2 = 2ma^2 V_0/\hbar^2, \quad \xi > 0, \eta > 0.$$

Hence find the conditions on V_0 so that there will be

(i) at least one bound s-state.
(ii) at least two bound s-states.
(iii) at least one bound p-state.

Place the two lowest s-states and the lowest p-state in order of increasing energy, and estimate the energy difference between the lowest s-states when $2ma^2 V_0/\hbar^2 \gg 1$.

FURTHER READING

1. MARGENAU and MURPHY, *The Mathematics of Physics and Chemistry*, van Nostrand, N.Y., 1956; for the Legendre and Laguerre functions and also the Hermite functions in Ch. 3.

2. MESSIAH, A., *Quantum Mechanics*, North Holland, 1962, vols. 1 and 2; especially vol. 2 for symmetries and angular momentum.

CHAPTER 5

Atoms and Molecules

5.1 Many Particles

Up to now we have only considered the motion of a single particle, except for a brief discussion of the classical motion of a rigid body in §1.4; even when we considered the hydrogen atom we reduced it to the motion of a single electron around a static nucleus. We now want to consider how we may extend the ideas of quantum mechanics to the motion of systems containing two or more particles. These particles may be identical ones, as is the case for the motion of a number of electrons round a heavy nucleus, or they may be distinct, as is the case for the electron and proton in the hydrogen atom. In particular we want to consider the idea of a wave function describing a system of more than one particle.

Let us consider in detail a system of two particles. Each of these particles, if it were on its own, could be described by a wave function; if $\mathbf{r}_1$ and $\mathbf{r}_2$ are used to denote the positions of the two particles, these wave functions would be $\psi_1(\mathbf{r}_1)$ and $\psi_2(\mathbf{r}_2)$. If the particles do not interact in any fashion, then we expect that the probability of finding the first particle in a small volume $d^3\mathbf{r}_1$ about the point $\mathbf{r}_1$, *and* of finding the second particle in a small volume $d^3\mathbf{r}_2$ about the point $\mathbf{r}_2$, is just the product of these separate probabilities; by the probability interpretation introduced in §2.5 this will be

$$|\psi_1(\mathbf{r}_1)|^2 \, d^3\mathbf{r}_1 |\psi_2(\mathbf{r}_2)|^2 \, d^3\mathbf{r}_2 \tag{5.1}$$

This result could be obtained by considering the system of two particles to be described by the product wave function

$\psi_{12}(\mathbf{r}_1, \mathbf{r}_2) = \psi_1(\mathbf{r}_1)\psi_2(\mathbf{r}_2)$. The interpretation for this wave function ψ_{12}, according to (5.1), is that $|\psi_{12}(\mathbf{r}_1,\mathbf{r}_2)|^2 \, d^3\mathbf{r}_1 \, d^3\mathbf{r}_2$ is the probability of finding the first particle in a small volume $d^3\mathbf{r}_1$ at $\mathbf{r}_1$, and the second particle in a small volume $d^3\mathbf{r}_2$ at $\mathbf{r}_2$. For non-interacting particles the total energy of the particles is just the sum of the individual energies of the separate particles. If the two particles have momenta and masses $\mathbf{p}_1, \mathbf{p}_2$ and m_1, m_2 respectively, then the total kinetic energy is

$$E = (\mathbf{p}_1^2/2m_1) + (\mathbf{p}_2^2/2m_2) \tag{5.2}$$

If there is not only no interaction between the particles, but also no external force acting on the particles, then E of (5.2) will be the total energy of the particles. Let us suppose that the separate particles have energy E_1 and E_2 respectively, so that $E = E_1 + E_2$ where $E_1 = \mathbf{p}_1^2/2m_1$, $E_2 = \mathbf{p}_2^2/2m_2$. We know that the time dependent wave functions which describe each of the particles has the form

$$\psi_1(\mathbf{r}_1, t) = e^{-iE_1 t/\hbar}\psi_1(\mathbf{r}_1), \quad \psi_2(\mathbf{r}_2, t) = e^{-iE_2 t/\hbar}\psi_2(\mathbf{r}_2) \tag{5.3}$$

so that the two-particle system will have the time dependent product wave function

$$\psi_{12}(\mathbf{r}_1, \mathbf{r}_2, t) = e^{-i(E_1 + E_2)t/\hbar}\psi_{12}(\mathbf{r}_1, \mathbf{r}_2) \tag{5.4}$$

The time dependence of this product wave function is that expected for a system with total energy $(E_1 + E_2)$, which is the correct total energy. We can now obtain the extension of the Schrödinger equation to this two particle system. We know that for the first particle the Schrödinger equation is

$$i\hbar \partial \psi_1(\mathbf{r}_1, t)/\partial t = H_1 \psi_1(\mathbf{r}_1, t) = E_1 \psi_1(\mathbf{r}_1, t) \tag{5.5}$$

In (5.5) $H_1 = \mathbf{p}_1^2/2m_1 = -\hbar^2 \nabla_1^2/2m_1$, where $\nabla_1 = (\partial/\partial x_1, \partial/\partial y_1, \partial/\partial z_1)$. Similarly

$$i\hbar \partial \psi_2(\mathbf{r}_2, t)/\partial t = H_2 \psi_2(\mathbf{r}_2, t) = E_2 \psi_2(\mathbf{r}_2, t) \tag{5.6}$$

We now perform the time differentiation of the time dependent, two-particle wave function $\psi_{12}(\mathbf{r}_1, \mathbf{r}_2, t)$ of (5.4):

$$\begin{aligned} i\hbar \partial \psi_{12}(\mathbf{r}_1, \mathbf{r}_2, t)/\partial t &= (E_1 + E_2)\psi_{12}(\mathbf{r}_1, \mathbf{r}_2, t) \\ &= E_1 \psi_1(\mathbf{r}_1, t)\,\psi_2(\mathbf{r}_2, t) + \psi_1(\mathbf{r}_1, t)\,E_2\psi_2(\mathbf{r}_2, t) \\ &= H_1 \psi_1(\mathbf{r}_1, t)\,\psi_2(\mathbf{r}_2, t) + \psi_1(\mathbf{r}_1, t)\,H_2\psi_2(\mathbf{r}_2, t) \\ &\qquad\qquad\qquad\qquad\qquad\qquad \text{(from (5.5), (5.6))} \\ &= (H_1 + H_2)\psi_{12}(\mathbf{r}_1, \mathbf{r}_2, t) \tag{5.7} \end{aligned}$$

The resulting extension of the Schrödinger equation is

$$i\hbar \frac{\partial}{\partial t} \psi_{12}(\mathbf{r}_1, \mathbf{r}_2, t) = (H_1 + H_2)\psi_{12}(\mathbf{r}_1, \mathbf{r}_2, t) \quad (5.8)$$

This is just the Schrödinger equation with Hamiltonian $(H_1 + H_2)$, as is to be expected, since $E = E_1 + E_2$, so the energy is additive in the separate particle energies.

We may now extend these results to the case of two non-interacting particles which are in some external field of force, for example two electrons moving round the helium nucleus (of charge equal to the negative of the charges on the two electrons, being composed of two protons and two neutrons). In this case the energies E_1 and E_2 of the separate particles will no longer be purely their kinetic energies, but will involve the effects of the potential energy V; the two particles will be described by a product wave function $\psi_{12}(\mathbf{r}_1, \mathbf{r}_2, t)$ of the form (5.4) where now (5.5) and (5.6) are still satisfied, though with

$$H_1 = -(\hbar^2 \nabla_1^2/2m_1) + V(\mathbf{r}_1), \qquad H_2 = -(\hbar^2 \nabla_2^2/2m_2) + V(\mathbf{r}_2)$$

so that the total Hamiltonian in the Schrödinger equation (5.8) is still $(H_1 + H_2)$. The probability interpretation associated with ψ_{12} is unchanged.

The final type of two-particle system we can consider is that for which the interaction between the two particles is taken into account, such as the two electrons moving in the helium nucleus, for which we consider the Coulomb repulsion between the electrons. In this case we cannot take a pure product wave function to describe the state of the two-particle system, nor can we express the energy of the classical motion as a sum of two terms, one depending only on the co-ordinate of the first particle, the other on the co-ordinate of the second particle. If $U(\mathbf{r}_{12})$ is the potential energy for the interaction between the particles, where $\mathbf{r}_{12} = \mathbf{r}_1 - \mathbf{r}_2$ is their separation vector, and they are again both moving in an external field of force with potential $V(\mathbf{r})$, their classical energy is

$$E = (\mathbf{p}_1^2/2m_1) + (\mathbf{p}_2^2/2m_2) + U(\mathbf{r}_{12}) + V(\mathbf{r}_1) + V(\mathbf{r}_2) \quad (5.9)$$

We extend the quantum mechanical discussion for the two-particle system which we have given above in as natural a fashion as possible. This is achieved by taking a wave function $\psi_{12}(\mathbf{r}_1, \mathbf{r}_2, t)$, to describe the state of the two particle system which is more general than the product wave function (5.4); it will have the same probability interpretation as the product case (5.4), and will satisfy the time dependent Schrödinger equation

$$i\hbar \partial \psi_{12}(\mathbf{r}_1, \mathbf{r}_2, t)/\partial t = H\psi_{12}(\mathbf{r}_1, \mathbf{r}_2, t) \qquad (5.10)$$

where H is obtained from the classical energy (5.9) by the natural replacements $\mathbf{p}_1 \to -i\hbar \nabla_1, \mathbf{p}_2 \to -i\hbar \nabla_2$ so that

$$H = \frac{-\hbar^2 \nabla_1^2}{2m_1} - \frac{\hbar^2 \nabla_2^2}{2m_2} + U(\mathbf{r}_{12}) + V(\mathbf{r}_1) + V(\mathbf{r}_2) \qquad (5.11)$$

Both the probability assumption and the Schrödinger equation (5.10) with (5.11) can only be justified by successful predictions obtained by means of these assumptions. Such successes are indeed present in ample evidence in atomic and molecular physics, and some of them will be described in this chapter and the next, but first we will briefly remark on the more general case of an N-particle system, where $N \geqslant 2$. From our previous discussion we expect the state of such a system to be described by the wave function $\psi(\mathbf{r}_1, \mathbf{r}_2, \cdots, \mathbf{r}_N, t)$ which depends on the separate particle position variables $\mathbf{r}_1, \mathbf{r}_2, \cdots, \mathbf{r}_N$. The probability interpretation for ψ is that

$$|\psi(\mathbf{r}_1, \mathbf{r}_2, \cdots, \mathbf{r}_N, t)|^2 d^3\mathbf{r}_1 d^3\mathbf{r}_2, \cdots, d^3\mathbf{r}_N$$

is the probability of finding simultaneously at time t the first particle in the volume $d^3\mathbf{r}_1$ at the point $\mathbf{r}_1$, the second in the volume $d^3\mathbf{r}_2$ at $\mathbf{r}_2$ and so on. The development with time of the N-particle wave function is described by means of the N-particle Schrödinger equation

$$i\hbar \partial \psi(\mathbf{r}_1, \cdots, \mathbf{r}_N, t)/\partial t = H\psi(\mathbf{r}_1, \cdots, \mathbf{r}_N, t) \qquad (5.12)$$

where the Hamiltonian H is obtained from the total classical Hamiltonian $H(\mathbf{r}_1, \cdots, \mathbf{r}_N, \mathbf{p}_1 \cdots \mathbf{p}_N)$ for the N-particle system by the usual quantum mechanical rule

$$H(\mathbf{r}_1, \cdots \mathbf{r}_N, \mathbf{p}_1, \cdots, \mathbf{p}_N) \to H(\mathbf{r}_1, \cdots, \mathbf{r}_N, -i\hbar \nabla_1, \cdots, -i\hbar \nabla_N)$$

In particular, for a system of N particles each moving in an external potential $V(\mathbf{r})$ at the point $\mathbf{r}$, and with a potential $U(\mathbf{r}_{ij})$ describing the interaction between the ith particle at $\mathbf{r}_i$ and the jth particle at $\mathbf{r}_j$, where $\mathbf{r}_{ij} = \mathbf{r}_i - \mathbf{r}_j$, then

$$H(\mathbf{r}_1, \cdots, \mathbf{r}_N, -i\hbar\nabla_1, \cdots, -i\hbar\nabla_N)$$

$$= \sum_{j=1}^{N}(-\hbar^2\nabla_j^2/2m_j) + \sum_{j=1}^{N} V(\mathbf{r}_j) + \sum_{i<j} U(\mathbf{r}_{ij}) \qquad (5.13)$$

The problem of finding the eigenvalues of such a Hamiltonian for different forms of V and U and different values of N will occupy us for the remainder of this chapter.

5.2 The Hydrogen Atom

The simplest example of a system containing more than one particle is the hydrogen atom, which consists of an electron of charge e and mass m moving round a proton of charge $-e$ and mass M, under the mutual Coulomb attraction between the two particles. The potential for this attraction is $-e^2/(4\pi\varepsilon_0|\mathbf{r}_{12}|)$, where $\mathbf{r}_1$ and $\mathbf{r}_2$ are the positions of the electron and proton, whilst $\mathbf{r}_{12} = \mathbf{r}_1 - \mathbf{r}_2$. The Hamiltonian for the hydrogen atom will then be

$$H = (\mathbf{p}_1^2/2m) + (\mathbf{p}_2^2/2M) - e^2/(4\pi\varepsilon_0|\mathbf{r}_{12}|) \qquad (5.14)$$

If we translate both the electron and the proton in the hydrogen atom through the same distance, then the same hydrogen atom is obtained, in other words the electron-proton system has a Hamiltonian (5.14) which is invariant under space translations (as can be seen by direct inspection of (5.14)). Such invariance leads to a conserved momentum, as we saw at the end of the previous chapter (§4.6); since the invariance is under translation of the *total* hydrogen atom we expect that the conserved momentum will be the *total* momentum of the hydrogen atom. We can see this most directly if we rewrite the Hamiltonian (5.14) in the centre of mass co-ordinates $\mathbf{R}$ and the relative co-ordinates $\mathbf{r}$, where

$$\mathbf{R} = (m\mathbf{r}_1 + M\mathbf{r}_2)/(m+M), \qquad \mathbf{r} = \mathbf{r}_1 - \mathbf{r}_2 \qquad (5.15)$$

so that

$$\mathbf{r}_1 = \mathbf{R} + \left(\frac{\mu}{m}\right)\mathbf{r}, \quad \mathbf{r}_2 = \mathbf{R} - \left(\frac{\mu}{m}\right)\mathbf{r}, \quad \mu = \frac{mM}{(m+M)} \qquad (5.16)$$

Using the chain rule of differentiation we then have

$$\left.\begin{array}{l}\nabla_1 = \{m/(m+M)\}\nabla_R + \nabla_r \\ \nabla_2 = \{M/(m+M)\}\nabla_R + \nabla_r\end{array}\right\} \quad (5.17)$$

where $\nabla_R = (\partial/\partial X, \partial/\partial Y, \partial/\partial Z)$ with $\mathbf{R} = (X, Y, Z)$, etc. If we substitute (5.17) into (5.14) we obtain, with $\mathbf{p}_1 = -i\hbar\nabla_1$, $\mathbf{p}_2 = -i\hbar\nabla_2$,

$$H = -\frac{\hbar^2}{2m}\left\{\left(\frac{m}{m+M}\right)\nabla_R + \nabla_r\right\}^2 - \frac{\hbar^2}{2M}\left\{\left(\frac{M}{M+m}\right)\nabla_R - \nabla_r\right\}^2 - \frac{e^2}{4\pi\varepsilon_0 r}$$

$$= -\frac{\hbar^2}{2(M+m)}\nabla_R^2 - \frac{\hbar^2}{2\mu}\nabla_r^2 - \frac{e^2}{4\pi\varepsilon_0 r} \quad (5.18)$$

We see that the form of H given by (5.18) involves the centre of mass co-ordinate $\mathbf{R}$ only in a kinetic energy term, so that the motion of the centre of mass is that of a free particle of mass $(m+M)$. We cannot determine this motion further, but have to assume a given value $\mathbf{P}$ for the centre of mass momentum, as given from some initial conditions. Then we may choose for the hydrogen atom wave function the separable form

$$\psi(\mathbf{r}_1, \mathbf{r}_2) = \psi(\mathbf{r}, \mathbf{R}) = e^{i\mathbf{P}\cdot\mathbf{R}/\hbar}\phi(\mathbf{r}) \quad (5.19)$$

so that if $H\psi(\mathbf{r}, \mathbf{R}) = E\psi(\mathbf{r}, \mathbf{R})$, then

$$\left(-\frac{\hbar^2}{2\mu}\nabla_r^2 - \frac{e^2}{4\pi\varepsilon_0 r}\right)\phi(\mathbf{r}) = \left\{E - \frac{\mathbf{P}^2}{2(m+M)}\right\}\phi(\mathbf{r}) \quad (5.20)$$

Thus (5.20) is a time independent Schrödinger equation in the variable $\mathbf{r}$; it corresponds to the motion of a particle of mass μ, called the reduced mass, in a Coulomb field from the origin, and with energy $\varepsilon = \{E - \mathbf{P}^2/2(m+M)\}$, which is the total energy of the hydrogen atom less the kinetic energy of the centre of mass motion. We evaluated the possible values of ε and the corresponding wave functions in §4.4, which treated the hydrogen atom as a one-particle problem. The values of ε were given by (4.61), though now with m_e replaced by μ, so that

$$\varepsilon_n = \frac{-\mu e^4}{2\hbar^2(4\pi\varepsilon_0)^2 n^2} = E_n\left(\frac{\mu}{m_e}\right) \quad (5.21)$$

where E_n is the value given by (4.61) with m_e as the electron's mass. Thus ε_n and E_N differ by the factor $\mu/m_e = M/(m+M)$, which is

1836/1837; the effect of the motion of the proton is small but noticeable. The total energy of the hydrogen atom is then

$$E = \varepsilon_n + \mathbf{P}^2/2(m+M)$$

Its wave function will be given by the product form (5.19) where $\phi(\mathbf{r})$ is the hydrogen atom wave function of (4.51) and (4.63).

We may also separate out the centre of mass into a similarly trivial kinetic energy term for a system of N particles described by the Hamiltonian (5.13) when it is invariant under space translations (so $V = 0$), but we will not do so here.

5.3 Perturbation Theory of Energy Levels

The motion of a system of two particles moving only under their mutual interaction can be reduced to their relative motion, and so effectively to the motion of a single particle. For more than two particles the reduced motion when the centre of mass motion has been factored out is still a problem in at least two (three-dimensional) relative co-ordinate variables. As such it is very difficult, if not impossible, to solve exactly. Even the relative motion in the two-particle case need not be as simple as the hydrogen atom problem. This means that approximate methods must be developed to obtain the energy levels of atoms or molecules, and also to obtain the manner in which scattering occurs. We will return to the problem of the scattering of two particles in the next chapter; in this section we will discuss an approximation procedure, known as perturbation theory, to obtain approximate values of the energy levels of a many particle system. Such approximation methods are, of course, also very useful for the motion of a single particle, except in the special soluble cases of which there are far too few.

Suppose we have a Hamiltonian operator H for a quantum mechanical system which is of form

$$H = H_0 + \lambda V_0 \qquad (5.22)$$

where λ is a constant. The term H_0 is assumed to involve the kinetic energy terms plus a part of the potential energy terms, whilst λV_0 is the remaining part of the potential energy terms. There are two basic requirements on the manner in which H is divided into H_0 and λV_0:

(i) The energy levels E_n and corresponding wave functions ϕ_n for H_0 can be obtained exactly.

(ii) λV_0 is small in comparison to H_0, so may be regarded as a *perturbation* on H_0. This will be so if the constant λ is much less than one. For example, for two electrons each moving in a central field of force described by the potential V, we may take H_0 to be the total energy plus the potential energy terms V, whilst λV_0 is the Coulomb potential due to their interaction:

$$H_0 = \frac{-\hbar^2 \nabla_1^2}{2m} - \frac{\hbar^2 \nabla_2^2}{2m} + V(\mathbf{r}_1) + V(\mathbf{r}_2)$$

$$\lambda V_0 = \frac{-e^2}{4\pi\varepsilon_0 |\mathbf{r}_1 - \mathbf{r}_2|}$$

(5.23)

We are interested in finding the energy levels of the perturbed Hamiltonian $(H_0 + \lambda V_0)$, and especially in finding the effect of the perturbing term λV_0. In other words we are interested in the perturbation δE of one of the energy levels, E_0, say, so that $E_0 + \delta E$ is the energy level of the perturbed Hamiltonian $(H_0 + \lambda V)$. Since we assume λ is small, then δE will be small, whilst the wave function ψ which is the eigenstate of $(H_0 + \lambda V_0)$ with energy E will be close to the unperturbed wave function ϕ_0:

$$\left. \begin{array}{l} \psi = \phi_0 + \delta\psi \\ E = E_0 + \delta E \end{array} \right\}$$

(5.24)

with both $\delta\psi$ and δE small. But we know that

$$(H_0 + \lambda V_0)\psi = E\psi$$

or

$$(H_0 + \lambda V_0)(\phi_0 + \delta\psi) = (E_0 + \delta E)(\phi_0 + \delta\psi)$$

If we use the fact that $H_0 \phi_0 = E_0 \phi_0$ then this becomes

$$\lambda V_0 \phi_0 + H_0 \delta\psi + \lambda V_0 \delta\psi = \delta E \phi_0 + E_0 \delta\psi + \delta E \delta\psi \qquad (5.25)$$

In (5.25) we may neglect the terms $\lambda V_0 \delta\psi$ and $\delta E \delta\psi$ since they are of the *second order* of smallness in comparison to the other terms. We proceed to obtain δE and $\delta\psi$; we may in general express $\delta\psi$ as a linear combination of the eigenfunctions ϕ_n belonging to the unperturbed Hamiltonian

$$\delta\psi = \sum_{n\neq 0} a_n \phi_n \qquad (5.26)$$

We do not include the term $n = 0$ in (5.26) since the total function ψ in (5.24) contains ϕ_0, and we can assume that ψ is so normalized that the coefficient of ϕ_0 is taken to be one in it; $\delta\psi$ will then not contain ϕ_0 at all. If we substitute (5.26) into (5.25) and use that $H_0\phi_n = E_n\phi_n$ we have (dropping the second order smallness terms):

$$\lambda V_0\phi_0 + \sum_{n\neq 0} a_n E_n \phi_n = \delta E \phi_0 + E_0 \sum_{n\neq 0} a_n \phi_n \qquad (5.27)$$

We now assume that the wave functions ϕ_n are orthogonal to each other and normalized to unity, so that $(\phi_n, \phi_m) = \delta_{nm}$. Then if we take the inner product of both sides of (5.27) with ϕ_0, we have

$$\delta E = \lambda(\phi_0, V_0\phi_0) \qquad (5.28)$$

whilst taking the inner product with $\phi_r (r \neq 0)$, we have

$$\lambda(\phi_r, V_0\phi_0) + a_r E_r = E_0 a_r$$

or

$$a_r = \lambda(\phi_r, V_0\phi_0)/(E_0 - E_r) \qquad (5.29)$$

We see that (5.28) gives the energy shift δE in terms of λV_0 and the unperturbed wave function ϕ_0; the perturbed part $\delta\psi$ of the wave function is given by (5.29). These results are known as *first order perturbation theory* results, since we have neglected the terms in the eigenvalue equation (5.25) which are of second order in the small parameter λ; higher order perturbation terms can be obtained, and are indeed very important in understanding physical processes for sub-nuclear particles.

5.4 Time Dependent Perturbation Theory

A perturbation of the Hamiltonian may not only be regarded as changing energy levels of the unperturbed system, but also as causing a transition from one such energy level to another. Such perturbations are very important when we consider the emission of radiation from an atom when an electron transfers from one energy level to a lower one (or absorption on transition to a higher energy level). We wish to find the probability of the occurrence of such a transition. We will consider this problem in the same fashion that we did for the energy level shift, that is by perturbation theory. We take the total Hamiltonian again to be $(H_0 + \lambda V_0)$, as given by (5.22). Now we want to consider the time *dependent* Schrödinger equation for the time dependent wave function ψ:

$$i\hbar \partial\psi/\partial t = (H_0 + \lambda V_0)\psi \tag{5.30}$$

We suppose that the system is described by the unperturbed wave function ϕ_0 at time $t = 0$. At a general time t we can expand the wave function ψ as in (5.26), extending that expression to cover also $n = 0$ and time dependent coefficients $a_n(t)$:

$$\psi(t) = \sum_{n \geqslant 0} a_n(t)\phi_n e^{-iE_n t/\hbar} \tag{5.31}$$

We have also included the explicit time dependence $e^{-iE_n t/\hbar}$ multiplying each of the unperturbed eigenstates ϕ_n. Our condition at $t = 0$ is

$$\psi(0) = \phi_0$$

which requires that

$$\left.\begin{aligned} a_n(0) &= 0 \quad (n \neq 0) \\ &= 1 \quad (n = 0) \end{aligned}\right\} \tag{5.32}$$

If we assume that (5.31) is the solution of the Schrödinger equation (5.30) for positive times, then we require

$$i\hbar \sum_{n \geqslant 0} \left(\frac{da_n}{dt} \phi_n e^{-iE_n t/\hbar} + a_n E_n \phi_n e^{-iE_n t/\hbar} \right) = \sum_{n \geqslant 0} a_n(t)[E_n + \lambda V_0]\phi_n e^{-iE_n t/\hbar}$$

where we have used on the right-hand side of the above equation that $H_0 \phi_n = E_n \phi_n$. This latter equation becomes

$$i\hbar \sum_n \frac{da_n}{dt} \phi_n e^{-iE_n t/\hbar} = \lambda \sum a_n V_0 \phi_n e^{-iE_n t/\hbar} \tag{5.33}$$

If we now take the inner product of both sides of (5.33) with ϕ_j, and use that $(\phi_j, \phi_n) = 0$ unless $j = n$, $(\phi_j, \phi_j) = 1$, then we obtain

$$i\hbar \frac{da_j}{dt} e^{-iE_j t/\hbar} = \lambda \sum_n a_n(\phi_j, V_0 \phi_n)e^{-iE_n t/\hbar} \tag{5.34}$$

If λ is small the coefficients a_n for $n \neq 0$ are expected to remain small over appreciable times, so that we need only take the term $n = 0$ in

ATOMS AND MOLECULES

the summation on the right hand side of (5.34), with $a_0 = 1$. Then we obtain, for $j \neq 0$,

$$i\hbar \frac{da_j}{dt} = \lambda(\phi_j, V_0\phi_0)e^{i(E_j-E_0)t/\hbar}$$

whose solution is

$$a_j(t) = \left(\frac{\lambda}{i\hbar}\right)\int_0^t dt(\phi_j, V_0\phi_0)e^{i(E_j-E_0)t/\hbar} \quad (5.35)$$

This solution gives an approximate form of wave function $\psi(t)$, valid for times over which the perturbation λV_0 causes only small amounts of the extra wave functions ϕ_j ($j \neq 0$) to be added to the unperturbed wave function ϕ_0; in other words over times t so that $|a_j(t)| \ll 1$, for $j \neq 0$. The interpretation of $a_j(t)$ follows from the statement that the overlap or inner product between $\psi(t)$ and $\phi_n e^{-iE_n t/\hbar}$ determines the probability that the state described by $\psi(t)$ at time t has both energy E_n and the wave function $\phi_n e^{-iE_n t/\hbar}$; this probability is the squared modulus of this overlap (see §2.5). From (5.31) this overlap is just $a_n(t)$, so that the probability of the transition from the state ϕ_0 at time $t = 0$ to the state ϕ_n at time t is $|a_n(t)|^2$, where $a_n(t)$ is given approximately by (5.35).

Let us consider this transition probability in two special cases:

(i) V_0 is independent of the time t. Then we may evaluate the t-integral in (5.35) to give

$$a_j(t) = -(\phi_j, \lambda V_0\phi_0)(e^{i(E_j-E_0)t/\hbar}-1)/(E_j-E_0)$$

so that the transition probability for transition to the state ϕ_j is

$$|a_j(t)|^2 = 2|(\phi_j, \lambda V_0\phi_0)|^2\left\{1-\cos\frac{(E_j-E_0)t}{\hbar}\right\}\Big/(E_j-E_0)^2 \quad (5.36)$$

If $E_j \neq E_0$ then the transition probability will remain small at *all* times, the time dependence being a bounded oscillation due to the cosine factor. However, a transition probability which *increases* with time will arise from (5.36) for E_j near E_0, since then, using that $\cos x \sim 1 - \tfrac{1}{2}x^2$ as $x \sim 0$,

$$|a_j(t)|^2 \sim |(\phi_j, \lambda V_0\phi_0)|^2(t^2/\hbar^2) \quad (5.37)$$

We may handle this case in general by supposing that there are many final states with energy E_j near E_0; if the number of states per unit energy is $\rho(E_j)$, so that in the energy range E to $E+dE$ there are $\rho(E)dE$ states, we may determine the probability of a transition to any one of these states with energy near E_0. To do this we need to multiply the transition probability to the state with energy E_j by the number of states in a unit energy interval, and integrate over the energy E_j; we know that only the states with energy E_0 will actually give an appreciable transition probability, as we see from (5.37). Thus the transition probability to some state will be

$$2|(\phi_j, \lambda V_0 \phi_0)|^2 \int_{-\infty}^{+\infty} \{1 - \cos(E_j - E_0)t/\hbar\} \rho(E_j) \, dE_j / (E_j - E_0)^2$$

We may change variables, $(E_j - E_0)t/\hbar = x$, to give the transition probability

$$2|(\phi_j, \lambda V_0 \phi_0)|^2 \int_{-\infty}^{+\infty} \rho\{E_0 + (\hbar x/t)\}(t/\hbar)(1 - \cos x) \, dx/x^2$$

We expect that only large contributions come from x near zero, so we replace the density of states' function $\rho\{E_0 + (\hbar x/t)\}$ by its value at $x = 0$, $\rho(E_0)$; we also have that

$$\int_{-\infty}^{+\infty} dx(1 - \cos x)/x^2 = \pi$$

so that the transition probability is

$$\frac{2\pi}{\hbar} |(\phi_j, \lambda V_0 \phi_0)|^2 t \, \rho(E_0)$$

and the transition probability *per unit time* is

$$\frac{2\pi}{\hbar} |(\phi_j, \lambda V_0 \phi_0)|^2 \rho(E_0) \qquad (5.38)$$

This result is very simple, and also very important. It states that the transition probability per second, from the initial state ϕ_0 to a final state ϕ_j, is $(2\pi/\hbar)$ times the modulus squared of the element of the perturbing potential, λV_0, between these states multiplied by the density of states with energy E_0.

(ii) The second case is when λV_0 has a purely oscillatory time dependence:

$$\lambda V_0 = V_0 \cos 2\pi \omega t \qquad (5.39)$$

In this case the result of evaluating $a_j(t)$ by (5.35) gives

$$a_j(t) = -\frac{(\phi_j, V_0\phi_0)}{2\hbar} \left\{ \frac{e^{i(2\pi\omega+(E_j-E_0)/\hbar)t}-1}{(2\pi\omega+(E_j-E_0)/\hbar)} + \frac{e^{i(-2\pi\omega+(E_j-E_0)/\hbar)t}-1}{(-2\pi\omega+(E_j-E_0)/\hbar)} \right\} \tag{5.40}$$

As in the time independent case (i) which we examined above, the transition probability $|a_j(t)|^2$ will be negligible if $\hbar\omega \neq \pm(E_j-E_0)$. If $\hbar\omega = E_j-E_0$, say, then we may proceed as in case (i) by multiplying the transition probability to a particular state ϕ_j, $|a_j(t)|^2$ by the number of states per unit energy and integrating over the energies E_j; by exactly similar methods the result for the transition probability *per unit time* will be

$$\frac{2\pi}{\hbar} |(\phi_j, V_0\phi_0)|^2 \rho(E_0+\hbar\omega) \tag{5.41}$$

This will correspond to a transition from a state of energy E_j to one of energy E_0 with emission of energy in an oscillatory form with frequency ω given by the Bohr frequency condition, $\omega = (E_j-E_0)/h$. The second term in the bracket in (5.40) will give a transition probability per unit time in which ω is replaced by $-\omega$ in (5.41), so corresponds to transition by absorption of energy of frequency ω. We will use both the time independent and time dependent perturbation theory to discuss the Zeeman effect in the next section.

5.5 The Zeeman Effect

As an example of the first order energy shift δE of (5.28) let us consider an electron moving in a central field of force with potential $-Ze^2/(4\pi\varepsilon_0 r)$; we will determine the shift of the lowest energy level due to the change of Z to $(Z+1)$. Then

$$\left. \begin{array}{l} H_0 = (-\hbar^2\nabla^2/2m) - (Ze^2/4\pi\varepsilon_0 r) \\ \lambda V_0 = -e^2/(4\pi\varepsilon_0 r) \end{array} \right\} \tag{5.42}$$

We may obtain the wave function with lowest energy for H_0 by replacing e^2 by Ze^2 in the discussion of the hydrogen atom in §4.4, so that

$$E_0 = \frac{-me^4 Z^2}{2\hbar^2(4\pi\varepsilon_0)^2}, \qquad \phi_0 = ce^{-Zr/a}, \qquad a = \frac{4\pi\varepsilon_0\hbar^2}{me^2} \tag{5.43}$$

where c is a constant which has been chosen so as to normalize ϕ_0, so that $c^2 \int_0^\infty e^{-2Zr/a} r^2 dr = (1/4\pi)$. Then by (5.28) and (5.42),

$$\delta E = \frac{-\dfrac{1}{4\pi}\int_0^\infty (e^2/r)e^{-2Zr/a}r^2 dr}{\int_0^\infty e^{-2Zr/a}r^2 dr} = -\frac{mZe^4}{(4\pi\varepsilon_0)^2\hbar^2} \qquad (5.44)$$

This is to be compared with the exact value of δE obtained from (5.43):

$$\delta E = -me^4 \frac{\{(Z+1)^2 - Z^2\}}{2\hbar^2(4\pi\varepsilon_0)^2} = -\frac{me^4(Z+\tfrac{1}{2})}{(4\pi\varepsilon_0)^2\hbar^2} \qquad (5.44')$$

Evidently the error involved in using (5.44) in place of (5.44') is small if Z is large; this corresponds to λV_0 being small compared to H_0, as we see from (5.42).

As a second example we consider the effect of a weak magnetic field on the energy levels of an electron moving in a central field. We consider the magnetic field **B** described by the vector potential **A**, so that $\mathbf{B} = \nabla \wedge \mathbf{A}$. The classical Hamiltonian H for a charged particle of mass m_e moving in such a field was found in example 3 in §1.2. For an electron of charge $-e$, we have

$$H = \left(\frac{1}{2m_e}\right)(\mathbf{p} - e\mathbf{A})^2 + e\phi \qquad (5.45)$$

In (5.45) $-e\phi$ is the potential of the electron in the absence of the magnetic field **B**. The time dependent Schrödinger equation arising from (5.45) will be

$$i\hbar\frac{\partial \psi}{\partial t} = H\psi = \left\{\frac{(-i\hbar\nabla + e\mathbf{A})^2}{2m_e} - e\phi\right\}\psi$$

We may write the quadratic term in the square brackets as

$$(-i\hbar\nabla + e\mathbf{A})^2\psi = -\hbar^2\nabla^2\psi - i\hbar\nabla\cdot(e\mathbf{A}\psi) - i\hbar e\mathbf{A}\cdot\nabla\psi + e^2 A^2\psi$$

$$= -\hbar^2\nabla^2\psi - 2i\hbar e\mathbf{A}\cdot\nabla\psi - i\hbar e\psi(\nabla\cdot\mathbf{A}) + e^2 A^2\psi$$

We expect that the term involving e^2 can be neglected, since it involves the square of the vector potential and, in a weak magnetic field, will be much smaller than the other terms; also $\nabla\cdot\mathbf{A} = 0$. Thus

$$H\psi = \left(\frac{-\hbar^2\nabla^2}{2m_e} - e\phi\right)\psi + \frac{e\hbar}{2m_e i}\mathbf{A}\cdot\nabla\psi \qquad (5.46)$$

ATOMS AND MOLECULES

We have split our Hamiltonian into two terms,

$$H_0 = \frac{-\hbar^2 \nabla^2}{2m_e} - e\phi, \qquad \lambda V_0 = \frac{e\hbar}{2m_e i} \mathbf{A} \cdot \nabla \qquad (5.47)$$

to which we may now apply perturbation theory. We will do this for the case that $\phi = \phi(|\mathbf{r}|)$, so that the perturbed potential is spherically symmetric; further, we will only take the case of a constant magnetic field $\mathbf{B} = (0, 0, B)$, so that

$$\mathbf{A} = (-\tfrac{1}{2}By, \tfrac{1}{2}Bx, 0)$$

and the perturbing potential is

$$\lambda V_0 = \frac{e\hbar}{2m_e i} B\left(x\frac{\partial}{\partial y} - y\frac{\partial}{\partial x}\right) = \frac{eB}{2m_e} L_z \qquad (5.48)$$

The unperturbed eigenfunctions for the spherically symmetric potential were discussed in §4.3, and have the form

$$\phi_{nlm} = R_{nl}(r) P_l^m(\cos\theta) e^{im\phi} \qquad (5.49)$$

where $L_z \phi_{nlm} = \hbar m \phi_{nlm}$. We know that there are wave functions of the form ϕ_{nlm} of (5.49) which have the same energy E_{nl}; these wave functions differ by the values of m, which takes one of the values $-l, -l+1, \cdots, l-1, l$. In order to consider the perturbation theory for such a case, we have to extend our previous discussion of the first order perturbation theory to this case of $(2l+1)$-fold degeneracy. Let us consider the simplest case of degeneracy, which is two-fold degenerate and not necessarily related to the $(2l+1)$-fold degeneracy noted above. There are two states, $\phi_0^{(1)}, \phi_0^{(2)}$, both with energy E_0 for the Hamiltonian H_0. Due to the perturbation λV_0 we expect that both these states will have their energies slightly displaced to $E_0 + \varepsilon_1$, $E_0 + \varepsilon_2$, the displaced levels being described by the new wave functions

$$\left.\begin{aligned}\psi_a &= a_1\phi_0^{(1)} + a_2\phi_0^{(1)} + \delta\psi_a = \phi_a + \delta\psi_a \\ \psi_b &= b_1\phi_0^{(1)} + b_2\phi_0^{(2)} + \delta\psi_b = \phi_b + \delta\psi_b\end{aligned}\right\} \qquad (5.50)$$

where $\delta\psi_a$ and $\delta\psi_b$ can be expanded into a linear sum of the other unperturbed eigenfunctions ϕ_n ($n \neq 0$). Then we have the eigenvalue condition:

$$\left.\begin{aligned}(H_0 + \lambda V_0 - E_0 - \varepsilon_1)(\phi_a + \delta\psi_a) &= 0 \\ (H_0 + \lambda V_0 - E_0 - \varepsilon_2)(\phi_b + \delta\psi_b) &= 0\end{aligned}\right\} \qquad (5.51)$$

or, dropping some quantities of second order of smallness:

$$\left.\begin{array}{l}(\lambda V_0-\varepsilon_1)\phi_a+(H_0-E_0)\delta\psi_a = 0 \\ (\lambda V_0-\varepsilon_2)\phi_b+(H_0-E_0)\delta\psi_b = 0\end{array}\right\} \quad (5.52)$$

If we take the inner product of the equations (5.52) with ϕ_b and ϕ_a respectively we obtain

$$(\phi_b, \lambda V_0 \phi_a) = 0 \quad (5.53)$$

whilst the inner product of (5.52) with ϕ_a and ϕ_b respectively gives

$$\varepsilon_1 = (\phi_a, \lambda V_0 \phi_a), \quad \varepsilon_2 = (\phi_b, \lambda V_0 \phi_b) \quad (5.54)$$

where ϕ_a and ϕ_b are taken to be orthogonal and normalized:

$$(\phi_a, \phi_b) = 0$$

$$(\phi_a, \phi_a) = (\phi_b, \phi_b) = 1$$

Thus if ϕ_a and ϕ_b are *chosen* so that (5.53) is true, then the energy shifts ε_1 and ε_2 are given by (5.54). We may extend this more generally: if there is a $(2l+1)$-fold degeneracy for an energy E_n of the unperturbed Hamiltonian H_0 and if the $(2l+1)$ unperturbed wave functions $\phi_{nl,1}, \cdots, \phi_{nl,2l+1}$ are chosen to be normalized to unity, orthogonal to each other and also orthogonal to each other when taken in the scalar product (5.53):

$$(\phi_{nl,r}, \lambda V_0 \phi_{nl,s}) = 0 \quad (r \neq s) \quad (5.55)$$

then the energy shifts are given by

$$\varepsilon_m = (\phi_{nlm}, \lambda V_0 \phi_{nlm}) \quad (5.56)$$

for $m = 1, 2, \cdots, (2l+1)$. In the case of λV_0 given by (5.48) and ϕ_{nlm} given by (5.49) the conditions of orthogonality, normalization and the set of conditions (5.54) are all satisfied, with $m = -l, \cdots, +l$. Then by (5.56) the energy shifts are

$$\varepsilon_m = \left(\phi_{nlm}, \frac{eB}{2m_e}L_z\phi_{nlm}\right) = \frac{e\hbar Bm}{2m_e} \quad (5.57)$$

Thus for each l the $(2l+1)$-fold degenerate energy level is split into $(2l+1)$ levels, with splitting given by (5.57).

There will thus be a change in the frequency of spectral lines due to this energy splitting; if a transition is from the unperturbed level

E_{nl} to $E_{n'l'}$, then the unperturbed frequency ω is $\omega = (E_{nl} - E_{n'l'})/h$; due to the energy shifts (5.57), there will be lines with slightly shifted frequencies with shift

$$\Delta\omega = \frac{eB}{4\pi m_e}\Delta m \qquad (5.58)$$

where Δm is the change in the value of m in the transition from a particular state ϕ_{nlm} to another state $\phi_{n'l'm'}$, $\Delta m = m - m'$. We may determine the allowed values of Δm by again using first order perturbation theory. To do this we again consider the perturbing potential λV_0 given by (5.47), where now $\mathbf{A}(\mathbf{r})$ describes a plane wave corresponding to the emission of a single photon; we take $\mathbf{A}(\mathbf{r})$ to be along the z-axis so that $\mathbf{A}(\mathbf{r}) = \mathbf{e}\, e^{i\mathbf{k}\cdot\mathbf{r}}$ where $\mathbf{e} = (0, 0, 1)$ and the wavelength λ of the emitted photon is $\lambda = 2\pi/k$. We now wish to apply formula (5.41) to the transition, the frequency ω being given by the Bohr frequency condition above. The matrix element of the perturbing potential causing this single photon emission is

$$(\phi_{nlm}, \mathbf{A}\cdot\nabla\phi_{n'l'm'}) = \int d^3\mathbf{r}\,\phi^*_{nlm}(\mathbf{r})\,e^{i\mathbf{k}\cdot\mathbf{r}}\,(\partial/\partial z)\phi_{n'l'm'}(\mathbf{r}) \qquad (5.59)$$

If the region over which the $\mathbf{r}$-integration is performed in (5.59) is much less than the wavelength λ, then the quantity $\mathbf{k}\cdot\mathbf{r}$ is less than $2\pi r/\lambda$, so is much less than one; we may then replace the exponential $e^{i\mathbf{k}\cdot\mathbf{r}}$ in (5.59) by unity. If $n' > n$, the radius r of effective integration in (5.59) will be of order $4\pi\varepsilon_0 n\hbar^2/me^2$ (for a hydrogen-like atom, with a single electron moving outside some 'closed' core), whilst the frequency ω of the photon emitted in the transition is (again for a hydrogen-like atom),

$$\omega = \frac{me^4}{4\pi\hbar^3(4\pi\varepsilon_0)^2}\left(\frac{1}{n^2} - \frac{1}{n'^2}\right), \qquad \lambda = \frac{c}{\omega}$$

so that we require

$$r \ll \lambda$$

or

$$\frac{4\pi\varepsilon_0 n\hbar^2}{me^2 c} \ll \frac{4\pi\hbar^3(4\pi\varepsilon_0)^2}{me^4}\left(\frac{1}{n^2} - \frac{1}{n'^2}\right)^{-1}$$

or

$$\frac{n}{4\pi\cdot 137}\left(\frac{1}{n^2} - \frac{1}{n'^2}\right) \ll 1 \qquad (5.60)$$

where we use in (5.60) that $e^2/(4\pi\varepsilon_0 \hbar c) = 1/137$. Evidently (5.60) is always valid, so we may take as a first approximation to the perturbing energy (5.59) the value

$$\int d^3\mathbf{r}\, \phi^*_{nlm}(\mathbf{r})\, (\partial/\partial z)\, \phi_{n'l'm'}(\mathbf{r}) \tag{5.61}$$

If we now evaluate the integral in (5.61) in spherical polar coordinates, the term involving ϕ in this integration may be written down exactly if we use (4.52) for the wave functions $\phi_{nlm}, \phi_{n'l'm'}$ and is:

$$\int_0^{2\pi} e^{i(m'-m)\phi} d\phi$$

This integral is evidently zero unless $m = m'$ (remember m and m' are integers). If we had chosen the potential $\mathbf{A}(\mathbf{r})$ of the emitted radiation to be along the x or y axes, and combine the terms as $x \pm iy = r\sin\theta e^{\pm i\phi}$ these give angular parts involving ϕ:

$$\int_0^{2\pi} e^{i(m'-m\pm 1)\phi} d\phi$$

which again is zero unless $m'-m = \pm 1$. By (5.41) the transition from the state with wave function ϕ_{nlm} to $\phi_{n'l'm'}$ can occur with non-zero probability only if there is a non-zero perturbation matrix element (5.59) causing it; such a transition is governed by the selection rule

$$\Delta m = m - m' = 0 \text{ or } \pm 1 \tag{5.62}$$

This means that $\Delta\omega$ of (5.58) can take three possible values,

$$\Delta\omega = 0 \quad \text{or} \quad \frac{\pm eB}{4\pi m_e} \tag{5.63}$$

Thus each spectral line with frequency ω is split into *three* lines when an external uniform magnetic field B is applied, the splitting being given by (5.63) (the polarization of the photon can also be obtained from this argument, though we will not discuss that further here). This splitting is called the Zeeman effect; since the splitting (5.63) doesn't contain h, it is to be expected that this result could be obtained by a classical argument; this is true, though again we will not discuss it further here.

It has been found that besides the *normal Zeeman effect* described by (5.63) there is an *anomalous Zeeman effect*, in which doublets appear, and not triplets, in the presence of an external magnetic field.

In particular if we consider the case when both states are s-wave states, $l = l' = 0$ (so $m = m' = 0$) then we must have $\Delta m = 0$, so that we expect no Zeeman splitting by (5.58). It is found that there is, even so, an anomalous Zeeman effect, with a doublet splitting. There is also an anomalous effect found in the motion of a beam of atoms perpendicular to a magnetic field **B**. If the field is in the z-direction, and is a function of z, then the splitting of the energy levels of a single electron in the atom will be proportional to Bm, so there will be a force in the z-direction proportional to $m\partial B/\partial z$. This will cause the beam to split into as many equal parts as there are possible values for m. In particular, for the hydrogen atom, which has a single electron, there will be no splitting if this single electron is in an s-state, with $l = 0$ and so $m = 0$. However, in the Stern-Gerlach experiment it was found that a beam of silver atoms was split into two parts, contrary to expectation (for the silver atom has 47 electrons, the outermost of which may be regarded as moving in an s-state round a central core of charge $+e$ comprising the nucleus and the inner 46 electrons, and the silver atom behaves like a hydrogen atom, at least as far as the outer electron is concerned). These results showed that an electron moving in an s-state had a two fold degeneracy; this degeneracy of the electrons' energy levels could be split by the application of an external magnetic field to produce two (closely) separated levels in a similar manner to the splitting of the $(2l+1)$ levels with a given value of the orbital angular momentum l and differing values of its z-component m. This new degeneracy was called *electron spin*, and the degree of freedom of the spin takes two values, either spin up or spin down with respect to some arbitrary axis. This intrinsic spin of the electron is extremely important in detailed discussion of electrons; here we will take account of it purely by adding an extra label α to the wave function ϕ_{nlm} for a single electron in a potential, so that now the wave function is ϕ_{nlm_α}, where α takes the value 1 or 2, 1 denoting an electron with spin up, 2 an electron with spin down. We will see how important this spin is in the next section, when we discuss the periodic table.

5.6 Atoms and the Periodic Table

An atom is composed of n electrons moving round a central nucleus which is small, massive, and has equal and opposite charge to the n

electrons; we will consider the nucleus as being fixed at the origin. We have already discussed the quantum mechanics of a system of n particles in §5.1. We want to continue that discussion here by considering the extra properties which result if the n particles are identical, as is the case for n electrons. If $\mathbf{r}_1, \cdots, \mathbf{r}_n$ are the n position co-ordinates of the particles, the Hamiltonian for the set of particles will depend on these position co-ordinates (and the corresponding partial differential operators representing the momenta which we consider as depending on the position co-ordinates) so that

$$H = H(\mathbf{r}_1, \cdots, \mathbf{r}_n)$$

Since the particles are identical, the Hamiltonian must be unchanged under any permutation of the n co-ordinates (otherwise the particles would be distinguishable if we follow their development with time):

$$H(\mathbf{r}_1, \cdots, \mathbf{r}_n) = H(\mathbf{r}_{i_1}, \cdots, \mathbf{r}_{i_n}) \tag{5.64}$$

where $(i_1, \cdots, i_n)$ is any permutation of $(1, \cdots, n)$. If the wave function $\psi(0)$ for the particles at time 0 is either symmetric or antisymmetric under interchange of any pair of particles, then the wave function $\psi(t) = e^{-iHt/\hbar}\psi(0)$ also has this symmetry or antisymmetry under exchange of any pairs of particle co-ordinates, for the time translation operator $e^{-iHt/\hbar}$ will not affect the symmetry. This means that the symmetry of the wave function is the same for all time, so can be discussed without reference to the time. Now it is found experimentally that all known particles have *either* symmetric *or* antisymmetric wave functions, and they are distinguished as follows:

(a) Particles with symmetric wave functions are known as bosons; such particles include photons, mesons.
(b) Particles with antisymmetric wave functions are known as fermions; such particles include electrons, protons, neutrons, etc.

The symmetry or antisymmetry of the wave function under general permutations must take the form:

$$\left. \begin{array}{l} \psi(\mathbf{r}_1, \cdots, \mathbf{r}_n) = +\psi(\mathbf{r}_{i_1}, \cdots, \mathbf{r}_{i_n}) \quad \text{(symmetric)} \\ \psi(\mathbf{r}_1, \cdots, \mathbf{r}_n) = \varepsilon\psi(\mathbf{r}_{i_1}, \cdots, \mathbf{r}_{i_n}) \quad \text{(antisymmetric)} \end{array} \right\} \tag{5.65}$$

where ε is the signature of the permutation from $(1, \cdots, n)$ to

$(i_1, \cdots, i_n)$ (equal to minus one raised to a power equal to the number of interchanges of pairs of variables required to go from the ordering $(1, \cdots, n)$ to $(i_1, \cdots, i_n)$. It has moreover been found experimentally that all known particles are either bosons or fermions.

Let us now turn to consider a set of n electrons moving round a central nucleus. We neglect the Coulomb repulsions between the separate electrons, though this may become important for large n. In this approximation the electrons are moving independently of each other, so that if we ignore the antisymmetry (5.65) we can take for their complete wave function a product of one particle wave functions:

$$\psi(\mathbf{r}_1, \cdots, \mathbf{r}_n) = \psi_1(\mathbf{r}_1) \cdots \psi_n(\mathbf{r}_n) \qquad (5.66)$$

In order to have an antisymmetric wave function, we just completely antisymmetrize the wave function (5.66) in the variables $\mathbf{r}_1, \cdots, \mathbf{r}_n$; the result is the determinental wave function

$$\psi(\mathbf{r}_1, \cdots, \mathbf{r}_n) = \frac{1}{\sqrt{n!}} \begin{vmatrix} \psi_1(\mathbf{r}_1) & \psi_1(\mathbf{r}_2) & \cdots & \psi_1(\mathbf{r}_n) \\ \psi_2(\mathbf{r}_1) & \psi_2(\mathbf{r}_2) & \cdots & \psi_2(\mathbf{r}_n) \\ & \cdots & & \\ & \cdots & & \\ \psi_n(\mathbf{r}_1) & \psi_n(\mathbf{r}_2) & \cdots & \psi_n(\mathbf{r}_n) \end{vmatrix} \qquad (5.67)$$

This is evidently antisymmetric under interchange of any pair of co-ordinates, since the interchange of two rows of a determinant produces a change in sign of the total determinant. We may consider the determinantal wave function (5.67) as describing a system of electrons, one of which is in the state with wave function ψ_1, one in $\psi_2, \cdots$, one in ψ_n; we cannot say which electron is in which state, as is consistent with the indistinguishability of the electrons.

We now see that if two electrons are in the same state, then two rows of the determinant (5.67) will be equal, so that it will vanish. But a vanishing wave function describes a non-existent state, so that we have deduced the *Pauli exclusion principle*: two electrons cannot be in the same state in an atom. It is the exclusion principle which will allow us to build up the periodic table of the elements.

To do this, let us consider the wave functions of the electrons in different atoms. To do this most conveniently we denote the determinantal wave function (5.67) just by a product $\psi_1 \cdots \psi_n$ of the separate one-electron wave functions, without their co-ordinate labels, the antisymmetrization resulting in a determinantal wave function also being understood. We also denote the wave function by the principle quantum number n and the orbital angular momentum l, so that two electrons both in states with $n = 1$, $l = 0$ (so s-waves) will be denoted by $(1s)(1s)$, which we naturally write as $(1s)^2$; similarly a $1s$ and a $2s$ electron will have a two electron wave function $(1s)(2s)$. We are now ready to give the wave functions of various atoms, starting from hydrogen. We let Z denote the number of protons in the atomic nucleus, so for a neutral atom it will also consist of Z electrons. In Table II we list the various electronic wave functions of the atoms with increasing Z. The important properties that are used in building up this table are:

(i) The Pauli exclusion principle only allows at most two electrons in a state with given n, l and m, the extra degree of freedom being the spin variable, taking two possible values. Thus a helium atom can have two $(1s)$ electrons, one having spin up, the other spin down.

(ii) The neutral atom has its electrons in their lowest possible energy levels, so that the lowest values of n are filled up first.

(iii) If all the states for a given n are filled, the electrons possess a certain degree of stability, forming a 'closed' shell. If a further electron is added to such a closed shell (increasing Z by one at the same time, of course, to keep the atom neutral) the resulting atom is like a hydrogen atom, in the sense that it has a single electron moving round a central 'core' of opposite charge. Such a hydrogen-like atom is sodium, with a single $3s$ electron and a central core of a nucleus with charge $11e$ and 10 electrons with 'core' wave function $(1s)^2(2s)^2(2p)^6$. The single electron is more easily removed from such an atom, forming a positive ion, Na^+. The converse type of atom is an atom with one electron needed to make up a closed shell such as chlorine, which needs one electron to make up the closed shell $(1s)^2(2s)^2(2p)^6$. It will tend to acquire such an electron to form a negative Cl^- ion; in the case of sodium chloride this extra electron can come from the sodium atom, the charged Na^+ and Cl^- ions being bound by the Coulomb attraction between them (this is a *polar* bond). We will discuss other methods of binding atoms to form molecules in the next section.

TABLE 2
Electronic wave functions for atoms in the periodic table up to potassium ($Z = 19$)

Symbol for Atom	H	He	Li	Be	B	C	N	O	F
Z	1	2	3	4	5	6	7	8	9
Electronic wave function	$(1s)$	$(1s)^2$	$(1s)^2(2s)$	$(1s)^2(2s)^2$	$(1s)^2(2s)^2(2p)$	$(1s)^2(2s)^2(2p)^2$	$(1s)^2(2s)^2(2p)^3$	$(1s)^2(2s)^2(2p)^4$	$(1s)^2(2s)^2(2p)^5$

Symbol for Atom	Ne	Na	Mg	Al	Si	P	S	Cl
Z	10	11	12	13	14	15	16	17
Electronic wave function	$(1s)^2(2s)^2(2p)^6$	$(1s)^2(2s)^2(2p)^6(3s)$	$(1s)^2(2s)^2(2p)^6(3s)^2$	$(1s)^2(2s)^2(2p)^6(3s)^2 3p$	$(1s)^2(2s)^2(2p)^6(3s)^2(3p)^2$	$(1s)^2(2s)^2(2p)^6(3s)^2(3p)^3$	$(1s)^2(2s)^2(2p)^6(3s)^2(3p)^4$	$(1s)^2(2s)^2(2p)^6(3s)^2(3p)^5$

Symbol for Atom	Ar	K
Z	18	19
Electronic wave function	$(1s)^2(2s)^2(2p)^6(3s)^2(3p)^6$	$(1s)^2(2s)^2(2p)^6(3s)^2(3p)^6(4s)$

We have already seen that property (i) follows from the Pauli exclusion principle. Property (ii) follows from the discussion in §4.3; it is not completely valid, however, since for potassium ($Z = 19$) the last electron does not begin to fill up the ($3d$) level but goes into the ($4s$) level. This can be shown to arise from the effects of the intrinsic potential barrier, which was mentioned in §4.3. When this potential barrier is taken into account in the Hamiltonian, it allows a ($4s$) electron to penetrate closer to the nucleus than a ($3d$) electron, and so have lower energy. Property (iii) follows from property (ii) in the sense that the electrons in the lower energy levels are relatively inert with respect to chemical reactions as compared to the outermost electrons (those in the highest energy levels); the reason that closed shells are the most stable of all configurations is related to the property that spherically symmetric states have lower energy than states which do not possess this spherical symmetry (the closed shell being spherically symmetric).

In order to obtain reasonably accurate energy levels and wave functions for many electron atoms, we may use a variety of methods. We have already discussed one of them, perturbation theory, in which the Coulomb repulsion energy is regarded as a perturbation on the energy of the electrons moving independently in the presence of the nucleus. An alternative approach is by the variational method, which we will briefly describe now. In order to calculate the ground state energy E_0 and wave function ϕ_0 for the Hamiltonian H, we form the expression $W = (\psi, H\psi)/(\psi, \psi)$. Then if we vary ψ, the minimum value of W will be E_0 and the wave function ψ for which this minimum value of W is achieved will be ϕ_0. We can see this by taking an expansion of ψ in eigenfunctions ϕ_n of H with $H\phi_n = E_n\phi_n, E_n \geqslant E_0$; if $\psi = \sum a_n \phi_n$ then

$$W = \sum E_n |a_n|^2 / \sum |a_n|^2$$

or

$$W - E_0 = \sum (E_n - E_0)|a_n|^2 / \sum |a_n|^2$$

Evidently the least value of this is achieved when only the term $n = 0$ in the sum in the numerator is present since otherwise $(W - E_0)$ is positive; this is so for $\psi = \phi_0$, as we stated. The variational principle is most useful if we can choose a form for ψ depending on only a few parameters and possessing the general properties needed for an eigenstate of the system; the minimum of W as a function of these parameters is then to be found.

As an example, we calculate the ground state of the hydrogen atom, with $H = -\hbar^2 \nabla^2/2m - e^2/r$. We take the trial wave function $\psi(\mathbf{r}) = e^{-ar}$, and then

$$\frac{(\psi, H\psi)}{(\psi, \psi)} = \frac{\hbar^2}{2m}\left(a^2 - \frac{2ame^2}{\hbar^2}\right)^2$$

$$= \frac{\hbar^2}{2m}\left(a - \frac{me^2}{\hbar^2}\right)^2 - \frac{me^4}{2\hbar^2}$$

which has a minimum for $a = me^2/\hbar^2$, with minimum value $-me^4/2\hbar^2$; this is the exact value of the energy of the ground state.

If the nuclear motions are taken into account as well as the electronic motions, it is possible to show by variational methods that there are long range forces *between* atoms varying as |distance between nuclei|$^{-7}$ these forces are called Van der Waals forces, acting between neutral atoms or molecules which have no overlap between their wave functions. The forces arise due to polarization of the electron clouds in the atoms.

Other forces also act between atoms, which bind them to produce molecules. We turn to discuss these forces now.

5.7 Molecules

We have already considered the polar or ionic bond which binds atoms (existing as ions or charged atoms); quantum mechanics does not affect the calculation of ionic forces, these being of a distinctly classical nature. The other important interatomic forces which are quantum mechanical arise from overlap forces which occur when atoms are close enough for their wave functions to overlap; suitable configurations of the overlap cause a reduction of the total energy of the atoms, so producing binding. There are, of course, repulsive forces which arise when closed shells of electrons overlap, such as in Na^+ ions; there are also metallic forces. We will not consider these here, but briefly discuss the nature of the overlap forces which bind atoms to produce molecules. These forces are the basis of modern chemistry.

We will assume that the largest force arises when the wave functions of the atoms have largest overlap (in this situation it may be shown by perturbation theory that the energy of the system is least). Let us consider in detail the water molecule H_2O. We neglect the nuclei of the atoms, and consider only the electrons outside closed shells. The hydrogen atom has a $(1s)$ electron in a spherically symmetric state. The oxygen atom has electronic wave function $(1s)^2(2s)^2(2p)^4$, so will have six electrons outside the $n = 1$ closed shell. The two $(2s)$ electrons also form a closed sub-shell (not quite as closed as the $(1s)$ electrons), so are not to be considered when we consider the binding in H_2O. Of the four $(2p)$ electrons we take two of them in $2p_z$ orbitals, with opposite spins, so they again form a closed sub-shell ($2p_z$ is the wave function with $m = 0$, so has form $\psi(2p_z) = (z/r)u_2(r)$, where $u_2(r)$ is the radial function); the other two $(2p)$ orbitals are $2p_x$ and $2p_y$, with

$$\psi(2p_x) = (x/r)u_2(r), \qquad \psi(2p_y) = (y/r)u_2(r)$$

Thus the electrons in the $2p_x$ and $2p_y$ orbitals can take part in molecular building. In order to obtain maximum overlap with the two $(1s)$ electronic wave functions in the hydrogen atoms, these atoms must have their nuclei along the x and y axes; the hydrogen-oxygen bonds will thus be at 90° to each other. Due to repulsion between the hydrogen nuclei and other reasons, this angle is increased to about 105°; the resulting distribution of charge is shown in Fig. 5.1.

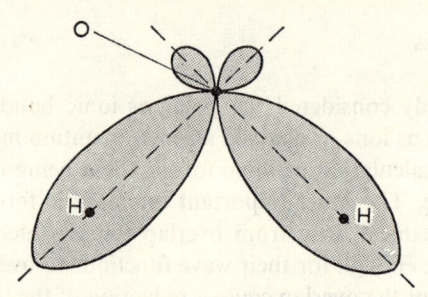

Fig. 5.1 The relative position of two hydrogen atoms and one oxygen atom to achieve maximum overlap of the $2p_x$ and $2p_y$ electrons of the oxygen atom with the 1s electrons of the two hydrogen atoms as shown.

Let us consider now the carbon atom which has two electrons with $n = 1$ and four with $n = 2$. The $n = 1$ electrons do not take part in chemical binding, so we have to consider electronic wave functions for four electrons formed from the basic wave functions $(2s)$, $(2p_x)$, $(2p_y)$, $(2p_z)$. We may take these wave functions to be linear combinations of the four basic wave functions; it is possible to choose these four wave functions so that the electrons are mainly on one or other of the lines from the corners of a tetrahedron to the centre (where the carbon nucleus is). These electronic wave functions, called tetrahedral orbitals, are of importance for example in the methane molecule CH_4, where they allow for maximum overlap with the electrons from each of four hydrogen atoms placed at the corners of the tetrahedron, as shown in Fig. 5.2.

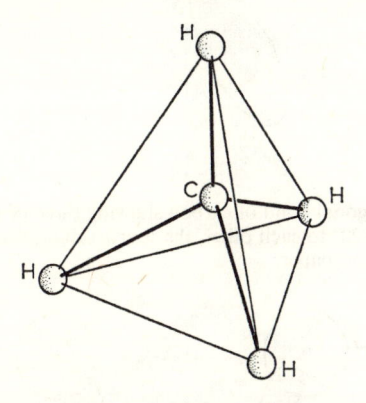

Fig. 5.2 The methane molecule CH_4, showing the directions of the tetrahedral orbitals of the carbon atom at the centre of a tetrahedron giving maximum overlap with the 1s electrons of the hydrogen atom at the vertices of this tetrahedron.

A molecule of great importance in organic chemistry is benzene, C_6H_6. Its structure may be understood in terms of the trigonal bond of carbon, which is obtained when three of the four electrons in the carbon atom lie mainly in one plane, and at 120° to each other; the fourth electron lies in a ($2p$) orbital of dumb-bell shape which is symmetrical on either side of the plane; this is shown in Fig. 5.2 (where only three of the electrons are shown). If carbon atoms are now set up in a ring, with pairs of electrons in the trigonal bonds in adjacent carbon atoms overlapping as much as possible, whilst the hydrogen atoms overlap with the other planar trigonal bond, a structure resembling that of Fig. 5.3 results. There remain the 6 electrons associated with the non-planar trigonal bonds; if these are shared equally between the carbon atoms in the ring, a remarkably stable structure results.

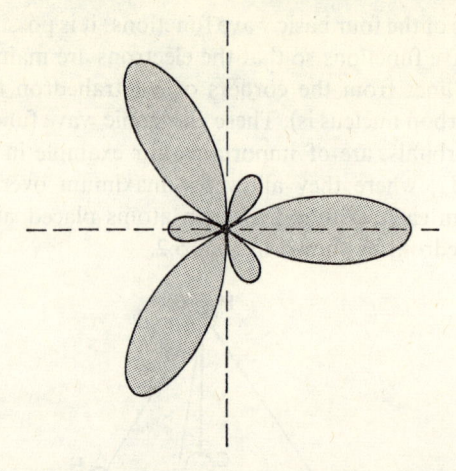

Fig. 5.3 The trigonal bond of carbon showing three of the four 2p electtron orbitals at 120° to each other, the fourth orbital being at right angles to the plane of the paper.

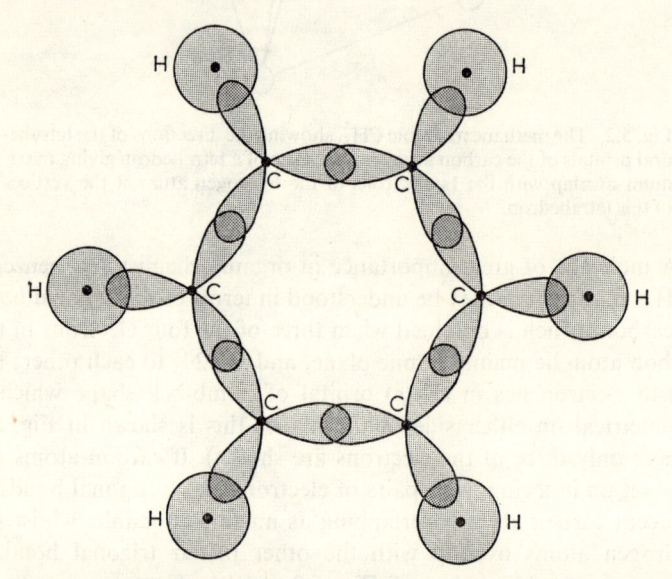

Fig. 5.4 The benzene ring C_6H_6, showing the overlap of three of the trigonal bonds of each carbon atom with two other carbon atoms and one hydrogen atom; the fourth orbital gives an electron which is shared among the carbon atoms of the ring, so there are 6 shared electrons; this sharing gives the remarkable stability of such components.

PROBLEMS

5.1 Show that the change in energy δE, for a system moving in one dimension and described by the Schrödinger equation $\psi'' + [E - U(x)]\psi = 0$, due to a small change δU in U, is given by

$$\delta E = \int \psi^*(x)\, \delta U(x)\, \psi(x)\, dx$$

(to first order).

5.2 If U is a harmonic oscillator potential and $\delta U = -Fx$, where F is a small constant, show that the first order change in energy levels is zero.

Give a classical argument that the shift of each energy level is $-F/2m\omega^2$ where the unperturbed $U(x) = \frac{1}{2}m\omega^2 x^2$.

(The notation is as in question 5.1.)

5.3 Let E_0 be the lowest energy level of a Hamiltonian H, and ψ be very close to the lowest eigenstate ϕ_0, $H\phi_0 = E_0\phi_0$. If $\psi = \phi_0 + \epsilon\phi$, where $\epsilon \ll 1$, expand the equation

$$E(\psi, \psi) - (\psi, H\psi) = 0$$

in powers of ϵ, where we take $E = E_0 + \epsilon E_1 + \epsilon^2 E_2$, dropping powers of ϵ above the second.

From the Hermitian property of H show that $E_1 = 0$, and by expanding ψ in terms of the eigenfunctions ϕ_n of H, with $H\phi_n = E_n\phi_n$, $E_n \geqslant E_0$, show that $E_2 \geqslant 0$. (A Hermitian operator is a self adjoint operator.)

What do you conclude from this about the value of $(\psi, H\psi)/(\psi, \psi)$?

5.4 Consider the single particle Hamiltonian

$$H = (p^2/2m) - (V_0\, e^{-r/r_0})/(r/r_0)$$

where V_0 and r_0 are constants. Use the trial wave function $\psi = e^{-\alpha r/r_0}$, with α a constant, to calculate $(\psi, H\psi)/(\psi, \psi)$.

What is the best approximation to the lowest energy level E_0 of H given by such a function. (Assume the result of question 5.3)? In particular show that this is given by

$$E = -(\alpha^2 \hbar^2/2mr_0^2)(2\alpha - 1)(2\alpha + 3)$$

where α satisfies

$$(2mV_0 r_0^2/\hbar^2)^2 = (2\alpha + 1)^3/2\alpha(2\alpha + 3)$$

5.5 Two particles, at $\mathbf{r}_1$ and $\mathbf{r}_2$, have Hamiltonian

$$H = (\mathbf{p}_1^2/2m_1) + (\mathbf{p}_2^2/2m_2) + V(|\mathbf{r}_1 - \mathbf{r}_2|)$$

Which of the quantities describing the linear and angular momentum for the system commute with H? What is the largest set of mutually commuting operators which can be chosen from this set?
(Hint: Work in the centre of mass and relative co-ordinates.)

5.6 A particle of mass m is free to move in two dimensions under the influence of the potential

$$V(x,y) = \epsilon \cos(\pi x/a) \cos(\pi y/a) \quad (0 < x < a, 0 < y < a)$$
$$= \infty \quad \text{(otherwise)}$$

where ϵ is small compared with $\hbar^2/ma^2$. Find the energy level of the ground state to first order in ϵ:

5.7 Three particles, with masses m_1, m_2 and m_3 respectively, have a total Hamiltonian (classically):

$$H = \mathbf{p}_1^2/2m_1 + \mathbf{p}_2^2/2m_2 + \mathbf{p}_3^2/2m_3 + V(\mathbf{r}_1 - \mathbf{r}_2)$$

Describe how the system may be quantized, and determine the energy levels in terms of the energy levels for various one-particle problems. What is the change in these levels if an additional term $V_3(\mathbf{r}_3)$ is added to H?

FURTHER READING

1. CONDON and SHORTLEY, *The Theory of Atomic Spectra*, C.U.P. 1953; the standard treatise on atomic spectra and the periodic table.
2. COULSON, C. A., *Valence*, Clarendon Press, 1952; a very readable account of the quantum theory of valency, the binding of atoms to form molecules.

CHAPTER 6

Scattering Theory

6.1 The Scattering Cross Section

Up to now we have been considering single particles or systems of particles, such as atoms or molecules, which are bound states. As we have seen, the wave functions which describe these bound states are localized in space so that the bound particle or particles are very close, on average, to the centre of force. Indeed, in our discussion of the hydrogen atom in §4.4, we rejected wave functions for the electron which caused the electron to reside mainly at a very large distance from the centre of force; such a rejection was only possible if the electron energy took one or other of a discrete set of possible values, which were the eigenvalues of the Hamiltonian operator, and all of which were negative. On the other hand, we have also considered the single particle wave function which is a plane wave; this evidently does not describe a localized particle, but a particle uniformly spread throughout space. If we consider this wave function as being used to describe a beam of particles, all with the same momentum, we can ask how such a beam is affected by a potential, $V(\mathbf{r})$.

We will assume that the potential is spherically symmetrical and decreases rapidly as we go away from the centre of force, vanishing at an infinite distance away from it. In this case, we are considering a beam of particles impinging on the potential and we expect this beam to be *scattered* by the potential. We expect the potential to distort the plane wave near the scattering centre, whilst far away such distortion can be neglected. If the plane wave is $e^{i\mathbf{p}\cdot\mathbf{r}/\hbar}$, so that the momentum of the particles is **p**, then, very far away from the scattering centre, the energy of each of the particles is $\mathbf{p}^2/2m$, where m is the particle mass. This is different from the case of the bound states of the electron in the hydrogen atom, when each electron had negative energy. In that case the classical allowed region is shown in Fig. 6.1; we also show

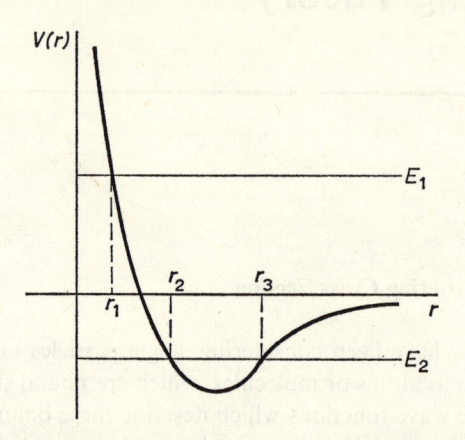

Fig. 6.1 The allowed regions of classical motion of an electron in a hydrogen atom potential when (a) the energy E_1 allows unlimited motion for $r > r_1$, and the electron is scattered (b) the energy E_2 gives a limited region of motion $r_2 < r < r_3$ and the electron is bound.

the region of allowed motion when the energy is positive. For negative energy, the electron is evidently localized, which is also the quantum mechanical result. For positive energy, the electron can move off to infinity, and when very far from the proton at the origin, it will behave like a free particle (we are neglecting complications which come from the difficulty of the long range of the Coulomb potential). In this positive energy case we expect the electron to be described by a superposition of plane waves or spherical waves in quantum mechanics; such waves can come in from a long distance away, be scattered, and go off to a long distance.

The process of the scattering of a particle by a centre of force plays as important a role in modern physics as does the existence of bound states. We may indeed regard the scattering process as giving another facet of the potential causing the scattering; the detailed knowledge of the scattering processes *and* the bound states for a potential may be used to specify the potential. Since in many situations the potential is not known, especially for the sub-nuclear particles, we have to find out some of the properties of the potential by observing the details of scattering caused by it.

How do we describe a scattering process in detail? We have already considered the initial stages. Experimentally, a beam of particles, each of momentum **p**, is directed onto the scattering centre; the setting up of such a beam is usually achieved by means of suitable collimators, which select particles moving in a given direction, together with bending magnetic fields (at least for charged particles). A collimator is then placed at an angle θ to the incoming beam direction, and the number of particles scattered into the collimator per unit time is measured; the set-up is shown in Fig. 6.2. We note that at least for

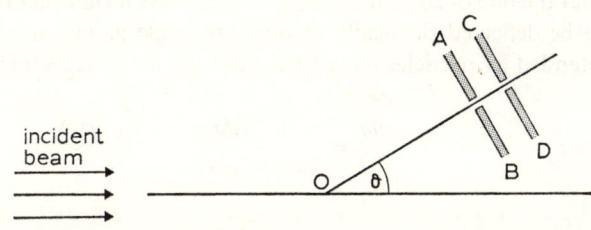

Fig. 6.2 The scattering of an incident beam by a field of force centred at O; the collimators AB, CD single out particles scattered through the angle θ with respect to the incident beam.

$\theta \neq 0$ there is no confusion between the incoming particles and those scattered through the angle θ, the former not being accepted by the collimator. Evidently the number of particles scattering into the observing apparatus will increase as we increase the aperture of the collimators. If we observe the particles scattered into an area dA normal to the direction from the centre of force, and at a distance r from the centre of force, or into a solid angle $d\Omega = dA/r^2$ (this being the definition of solid angle), we expect that the number N being scattered per second into $d\Omega$ will be proportional to $d\Omega$. Let us suppose further that we have one particle incident per unit area per unit time perpendicular to the path of the incident beam. For such a situation we define the constant of proportionality which, when multiplied by $d\Omega$, gives the number scattered per second into $d\Omega$, to be the *differential scattering cross section* $d\sigma/d\Omega$:

$$\frac{d\sigma}{d\Omega} = \frac{\text{number of particles scattered into } d\Omega \text{ at angle } \theta \text{ per unit time}}{(\text{number of particles incident per unit area per unit time}) \, d\Omega}$$

$$= \frac{N}{d\Omega} \qquad (6.1)$$

We see that the differential cross section has the dimensions of an area (which arises from the 'per unit area' in the denominator in (6.1)). Indeed $d\sigma$ may be interpreted as the area of a disc with centre at the scattering centre, which is placed perpendicular to the path of the incident beam, and which must be struck by this beam in order that the particles be scattered into the solid angle $d\Omega$ at the angle θ. We may relate this to the classical motion of the particle by means of the impact parameter, $b(\theta)$. This is the distance away from the centre of force that the line of the initial particle path must lie in order that the particle be deflected classically through an angle θ. The solid angle $d\Omega$ subtended by particles lying between θ and $\theta + d\theta$ is $2\pi \sin\theta \, d\theta$, so that

$$\frac{d\sigma}{d\Omega} = \frac{1}{2\pi \sin\theta} \frac{d\sigma}{d\theta} \qquad (6.2)$$

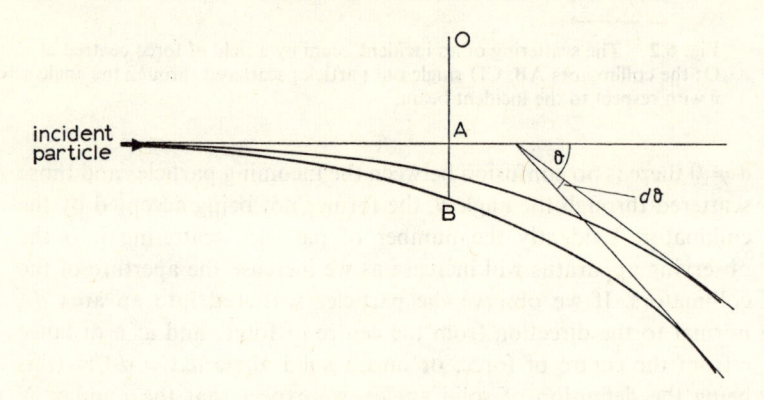

Fig. 6.3 The impact parameter $b(\theta)$ equal to AO; the change db in b is equal to AB in order that θ change to $\theta + d\theta$, as shown.

On the other hand, the small change $d\theta$ in the scattering angle will be obtained if b is changed to $b + db$ as shown in Fig. 6.3; this is achieved if the particle is moving in the ring of inner radius b and outer radius

$b + db$ perpendicular to its path. This ring has area $2\pi b\, db$, so that since $d\sigma$ is this area,

$$d\sigma = 2\pi b\, db \qquad (6.3)$$

From (6.2) and (6.3) the differential cross section is

$$\frac{d\sigma}{d\Omega} = \frac{1}{2\pi \sin\theta} \frac{d\sigma}{d\Omega} = \frac{b}{\sin\theta} \frac{db}{d\theta} \qquad (6.4)$$

Of course the result (6.4) only gives the classical result for $d\sigma$ if the classical value of $b(\theta)$ is used. This result may also be a good approximation to the quantum mechanical scattering, which is the case for the scattering of an electron by a fixed particle of charge Ze; classically the relation between θ and b is given by

$$mbv^2 \tan \tfrac{1}{2}\theta = Ze^2/4\pi\epsilon_0 \qquad (6.5)$$

where v is the particles velocity and m is its mass. Then from (6.4),

$$\frac{d\sigma}{d\Omega} = \frac{1}{4}\left(\frac{Ze^2}{4\pi\epsilon_0 mv^2}\right)^2 \frac{1}{(\sin\tfrac{1}{2}\theta)^4} \qquad (6.6)$$

which is the classical scattering formula of Rutherford. This formula is also valid in quantum mechanics, as we will see later; since the Rutherford formula does not contain h, it is to be expected that the classical and quantum results agree. We notice that the majority of the scattering occurs at small values of θ.

For a general scattering process, an important quantity describing the scattering is defined to be the *total cross section*, σ; it is evidently obtained from the differential cross section by integrating the latter over all solid angles:

$$\sigma = \int \frac{d\sigma}{d\Omega} d\Omega \qquad (6.7)$$

We may interpret σ as the area of a disc (perpendicular to the incoming particle's path) which must be struck in order that the particle be scattered (through any angle θ). The total cross section σ evidently contains much less information than the values of the differential cross section $d\sigma/d\Omega$ evaluated for various values of θ; it is still a very important quantity, and indicates how strongly particles are scattered out of the incoming beam. The larger σ is, the more such particles are scattered out of the beam, and conversely.

6.2 The Scattering Amplitude

Now that we have defined the differential cross section $d\sigma/d\Omega$, we must determine how to evaluate it for a given potential. Such an evaluation will allow us to predict the dependence of $d\sigma/d\Omega$ on θ for a given potential; if such dependence is not in agreement with experimental results, we will change the potential V till agreement between prediction and experiment is obtained for $d\sigma/d\Omega$. In this way we may find the form of the unknown potential causing the scattering, if it is indeed unknown. We can also see that not only does $d\sigma/d\Omega$ depend on the angle θ of scattering but also on the energy $\mathbf{p}^2/2m$ or the momentum $\mathbf{p}$ of the incident particle.

In order to find this dependence we introduce the *scattering amplitude* as follows. We already remarked that the plane wave function which describes the incident beam of particles will be distorted by the potential. This distortion will take the form of an additional wave function which vanishes for large distance r from the centre of force. This extra wave function will be a spherical wave spreading out from the scattering centre, of form $e^{ipr/\hbar}/r$; the time dependent form of spherical wave is $e^{i(pr-Et)/\hbar}/r$, where $E = p^2/2m$, and evidently as time increases we must increase r so that the phase of the wave, $(pr-Et)/\hbar$, remains constant. The wave will be also affected by a factor $f(\theta)$ depending on the angle θ of scattering, though not on the azimuthal angle ϕ corresponding to rotations about the path of the incident particle, due to the spherical symmetry of the potential. Thus the wave function $\psi(\mathbf{r})$ will be, for large $r = |\mathbf{r}|$, of approximate form

$$\psi(\mathbf{r}) = e^{i\mathbf{p}\cdot\mathbf{r}/\hbar} + (e^{ipr/\hbar}/r) f(\theta) \tag{6.8}$$

The quantity $f(\theta)$ is defined to be the scattering amplitude. We can determine the differential cross section $d\sigma/d\Omega$ in terms of $f(\theta)$. To do this we note that the incident particle current S is along the direction of the path of the incident particles, and has value (see §2.7, equation (2.54))

$$S = \frac{\hbar}{m} \text{Im}\left(e^{-ipz/\hbar} \frac{d}{dz} e^{ipz/\hbar}\right) = \frac{p}{m} = v \tag{6.9}$$

where v is the velocity of the incoming particle, and we take the z-axis along the direction of the incoming beam. Since the incident

SCATTERING THEORY 175

beam has one particle per unit volume (since $|e^{i\mathbf{p}\cdot\mathbf{r}/\hbar}|^2 = 1$), there will be v particles passing across unit area placed perpendicular to the particle path in one second, so that v is the flux of incident particles. We now need to consider the number of particles which are scattered through an angle θ and detected as scattered particles. We may again use the probability current of (2.54), though now we wish to consider the component of this current along the radial direction. We wish to do this only for the scattered part of the total wave function (6.8), that is for the spherical wave with large r behaviour $e^{ipr/\hbar}f(\theta)/r$; the plane wave part $e^{i\mathbf{p}\cdot\mathbf{r}/\hbar}$ is not observed by means of the detection apparatus, since the collimators single out only particles scattered through an angle θ. The scattered particle current along the radial direction is, at least for large r,

$$S_r = \frac{\hbar}{m} \operatorname{Im}\left[e^{-ipr/\hbar}\frac{f^*(\theta)}{r} \frac{\partial}{\partial r}\left\{f(\theta)\frac{e^{ipr/\hbar}}{r}\right\}\right]$$

$$= \frac{v}{r^2}|f(\theta)|^2 + \text{(terms behaving like } \frac{1}{r^3} \text{ for large } r\text{)} \quad (6.10)$$

where the terms omitted from the right-hand side of (6.10) fall off like $1/r^3$ or faster for large r. The number of particles scattered into the solid angle $d\Omega$ at the angle θ in one second will be equal to $S_r dA$, where dA is the area which subtends the solid angle $d\Omega$, so that $dA = r^2 d\Omega$. When we use these results and the definition (6.1), we have that

$$\frac{d\sigma}{d\Omega} = \frac{v}{r^2}|f(\theta)|^2 \frac{dA}{v d\Omega} = |f(\theta)|^2 \quad (6.11)$$

and the total cross section of (6.7) will be

$$\sigma = \int |f(\theta)|^2 d\Omega = 2\pi \int_0^\pi \sin\theta |f(\theta)|^2 d\theta$$

In order, then, to solve the problem of determining the differential or total cross section for a given potential, both for their angular and energy dependences, we have to obtain the value of the scattering amplitude $f(\theta)$, both as a function of the scattering angle θ and of the energy $\tfrac{1}{2}mv^2$ of the incident particle. We turn to the problem in the next section.

6.3 The Scattering Equation

In this section we wish to obtain a more detailed description of the scattering process than we have so far. To do this we will attempt to solve the time independent Schrödinger equation for the wave function $\psi(\mathbf{r})$, whose form at large distances is given by (6.8). If we denote $\mathbf{p}/\hbar$ by $\mathbf{k}$, then the energy of the scattered particle is equal to its energy at great distance when it is free and moving with momentum $\mathbf{p}$. Thus it has energy $\mathbf{p}^2/2m$, so that the time independent Schrödinger equation will be

$$(\mathbf{p}^2/2m)\psi(\mathbf{r}) = -(\hbar^2/2m)\nabla^2\psi + V(\mathbf{r})\psi(\mathbf{r})$$

or, in terms of $\mathbf{k}$:

$$(\nabla^2 + k^2)\psi = (2m/\hbar^2)V(\mathbf{r})\psi(\mathbf{r}) \tag{6.12}$$

Taking a hint from (6.8) we introduce the scattered wave function $\psi_s(\mathbf{r})$ so that

$$\psi(\mathbf{r}) = e^{i\mathbf{k}\cdot\mathbf{r}} + \psi_s(\mathbf{r}) \tag{6.13}$$

Then the scattering amplitude $f(\theta)$ is defined from the form of $\psi_s(\mathbf{r})$ as $r \sim \infty$:

$$\psi_s(\mathbf{r}) \sim (e^{ikr}/r)f(\theta) \quad \text{as } r \sim \infty \tag{6.14}$$

We may derive an equation for $\psi_s(\mathbf{r})$ from (6.12):

$$(\nabla^2 + k^2)\psi_s = (2m/\hbar^2)V(\mathbf{r})\psi(\mathbf{r}) \tag{6.15}$$

Let us treat the right hand side of (6.15) as a given quantity; we may then solve (6.15) for ψ_s as follows. We consider the *Green's function* $G(\mathbf{r}, \mathbf{r}')$ which is defined as the solution of

$$(\nabla^2 + k^2)G(\mathbf{r}, \mathbf{r}') = \delta^3(\mathbf{r} - \mathbf{r}') \tag{6.16}$$

It is evident that $G(\mathbf{r}, \mathbf{r}')$ is a function of $(\mathbf{r} - \mathbf{r}')$ only; let us denote this function by $G(\mathbf{r} - \mathbf{r}')$. Then (6.16) becomes

$$(\nabla^2 + k^2)G = \delta^3(\mathbf{r}) \tag{6.17}$$

whilst the solution of (6.15) is

$$\psi_s(\mathbf{r}) = \int G(\mathbf{r} - \mathbf{r}')(2m/\hbar^2)V(\mathbf{r}')\psi(\mathbf{r}')\,d^3r' \tag{6.18}$$

since if we apply $(\nabla^2 + k^2)$ to both sides of (6.18) the right-hand side, on using (6.17), becomes $(2m/\hbar^2)\int d^3r'\,\delta^3(\mathbf{r} - \mathbf{r}')V(\mathbf{r}')\psi(\mathbf{r}')$ which is

SCATTERING THEORY

$(2m/\hbar^2) V(\mathbf{r})\psi(\mathbf{r})$, and so is equal to the right hand side of (6.15). Thus ψ_s defined by (6.18) is a solution of (6.15). We now want to show that the solution of (6.17), which will also give the asymptotic form (6.14) for $\psi_s(\mathbf{r})$, for large r, is

$$G(r) = -e^{ikr}/4\pi r \tag{6.19}$$

Indeed the function e^{ikr}/r of (6.19) satisfies for $\mathbf{r} \neq 0$ the equation

$$(\nabla^2 + \mathbf{k}^2)G(\mathbf{r}) = 0$$

The radial part of ∇^2 in spherical co-ordinates is $(1/r^2)(\partial/\partial r)(r^2 \partial/\partial r)$, and so for $r \neq 0$,

$$(\nabla^2 + \mathbf{k}^2)\frac{e^{ikr}}{r} = \frac{1}{r^2}\frac{d}{dr}\left(r^2 \frac{d}{dr}\right)\frac{e^{ikr}}{r} + \frac{k^2 e^{ikr}}{r}$$

$$= \frac{1}{r^2}\frac{d}{dr}(-e^{ikr} + ikre^{ikr}) + \frac{k^2 e^{ikr}}{r}$$

$$= \frac{1}{r^2}[-ike^{ikr} + ike^{ikr} - k^2 re^{ikr} + k^2 re^{ikr}]$$

$$= 0$$

Thus equation (6.17) is satisfied for $\mathbf{r} \neq 0$; we have to verify it for $\mathbf{r} = 0$, or equivalently to show that for *any* smooth function $f(\mathbf{r})$,

$$-\int f(\mathbf{r})(\nabla^2 + \mathbf{k}^2)(e^{ikr}/4\pi r) d^3r = f(0) \tag{6.20}$$

Since we know that the integrand on the left of (6.20) is zero unless $\mathbf{r} = 0$ we may take $f(\mathbf{r})$ to be $f(0)$ under the integral. Thus we need to show that

$$-\frac{1}{4\pi}\int d^3r(\nabla^2 + \mathbf{k}^2)\left(\frac{e^{ikr}}{r}\right) = 1 \tag{6.21}$$

where the integral is taken over a sphere $|\mathbf{r}| \leq a$ with a arbitrarily small. To show (6.21) we discuss the integral on the left of (6.21), as confined to the sphere $|\mathbf{r}| \leq a$, in two parts:

$$\left.\begin{array}{l}\text{(i)} \quad -\dfrac{k^2}{4\pi}\displaystyle\int_{r\leq a} d^3\mathbf{r}\left(\dfrac{e^{ikr}}{r}\right) = k^2 \int_0^a re^{ikr}\,dr = ikae^{ika} \\[2ex] \text{(ii)} \quad -\dfrac{1}{4\pi}\displaystyle\int_{r\leq a} d^3\mathbf{r}\cdot\nabla^2\left(\dfrac{e^{ikr}}{r}\right)\end{array}\right\} \tag{6.22}$$

We can use Green's theorem applied to the sphere centred at the origin and of radius a to rewrite this integral (ii) as

$$-\frac{1}{4\pi}\int_{r=a} d\mathbf{S}\cdot\nabla\left(\frac{e^{ikr}}{r}\right) \qquad (6.23)$$

where $dS = \sin\theta d\theta d\phi$ is an element of surface area on the sphere $r = a$ and $d\mathbf{S}$ has the magnitude dS and points outward along the unit normal. The value of the integral described by (6.23) is then

$$-\frac{r^2}{4\pi}\int_0^\pi \sin\theta d\theta \int_0^{2\pi} d\phi \frac{d}{dr}\left(\frac{e^{ikr}}{r}\right)\bigg|_{r=a} = ika + e^{ika} \qquad (6.24)$$

We see that the sum of (6.22) and (6.24) is just e^{ika}; for very small a the value of this quantity is one, so that since the left-hand side of (6.20) is equal to the sum of (6.22) and (6.24) we have proved (6.21). Hence

$$\psi_s(\mathbf{r}) = -\frac{2\pi m}{\hbar^2}\int \frac{e^{ik|\mathbf{r}-\mathbf{r}'|}}{|\mathbf{r}-\mathbf{r}'|} V(\mathbf{r}')\psi(\mathbf{r}')d^3\mathbf{r}' \qquad (6.25)$$

Finally, let us show that if we use this value for the scattered wave ψ_s we obtain its correct asymptotic behaviour for large r as a purely *outgoing* spherical wave as given by the second term on the right of (6.8). Such a wave has the time dependent form

$$\frac{1}{r}e^{ikr-(i\hbar k^2 t/2m)}$$

so that it preserves its phase if r increases as t increases. We have to consider the behaviour of the right-hand side of (6.25) for large $|\mathbf{r}|$, with $\mathbf{r}$ itself lying along a direction at an angle θ to the direction of the incident particle motion (which is along the vector $\mathbf{k}$). Let us call $\mathbf{k}'$ the vector with the same length as $\mathbf{k}$ but along the direction of $\mathbf{r}$; $\hbar \mathbf{k}'$ denotes the momentum of the outgoing scattered particle. If we assume $V(\mathbf{r}')$ is negligible if $|\mathbf{r}'|$ becomes larger than some finite amount then we need only consider a finite range of integration over $|\mathbf{r}'|$ on the right of (6.25). In this case when $|\mathbf{r}|$ is very large we may replace $|\mathbf{r}-\mathbf{r}'|$ in the denominator of the integrand in (6.25) by $|\mathbf{r}|$, whilst we have

$$|\mathbf{r}-\mathbf{r}'| = \sqrt{(\mathbf{r}-\mathbf{r}')^2} = (r^2 - 2\mathbf{r}\cdot\mathbf{r}' + r'^2)^{1/2}$$

$$\sim |\mathbf{r}|\left(1 - \frac{2\mathbf{r}\cdot\mathbf{r}'}{|\mathbf{r}|}\right)^{1/2} \sim |\mathbf{r}| - \frac{\mathbf{r}\cdot\mathbf{r}'}{|\mathbf{r}|} \quad \text{as } |\mathbf{r}| \sim \infty$$

Then
$$k|\mathbf{r}-\mathbf{r}'| \sim k\left(|\mathbf{r}| - \frac{\mathbf{r}\cdot\mathbf{r}'}{|\mathbf{r}|}\right)$$

$k\mathbf{r}/|\mathbf{r}|$ is a vector of length equal to that of $\mathbf{k}$ but along the direction of $\mathbf{r}$. By our definition this is equal to $\mathbf{k}'$:

$$k|\mathbf{r}-\mathbf{r}'| \sim kr - \mathbf{k}'\cdot\mathbf{r}'$$

so the right-hand side of (6.25) becomes for large r:

$$\psi_s(\mathbf{r}) \sim \frac{e^{ikr}}{r}\left(-\frac{2\pi m}{\hbar^2}\right)\int e^{-i\mathbf{k}'\cdot\mathbf{r}'} V(\mathbf{r}')\psi(\mathbf{r}')\, d^3\mathbf{r}' \quad (6.26)$$

We see from the right hand side of (6.26) that $\psi_s(\mathbf{r})$ becomes an outgoing spherical wave for large $|\mathbf{r}|$, and that the scattering amplitude $f(\theta)$ is

$$f(\theta) = -(2\pi m/\hbar^2)\int e^{-i\mathbf{k}'\cdot\mathbf{r}'} V(\mathbf{r}')\psi(\mathbf{r}')\, d^3\mathbf{r}' \quad (6.27)$$

This expression for the scattering amplitude is important, though it requires knowledge of the total wave function $\psi(\mathbf{r}')$. We can find an equation for this total wave function by using (6.25) to replace ψ_s on the right of (6.13), so giving

$$\psi(\mathbf{r}) = e^{i\mathbf{k}\cdot\mathbf{r}} - \frac{2\pi m}{\hbar^2}\int \frac{e^{ik|\mathbf{r}-\mathbf{r}'|}}{|\mathbf{r}-\mathbf{r}'|} V(\mathbf{r}')\psi(\mathbf{r}')\, d^3\mathbf{r}' \quad (6.28)$$

This equation is an *integral equation* for the wave function ψ, in that ψ enters on the right hand side of the equation under the integral; its value at many points of space is involved in this integral. We may interpret (6.28) in a very natural fashion: the total wave function $\psi(\mathbf{r})$ at the point $\mathbf{r}$ is composed of the sum of two terms. The first of these is just the contribution from the incident particle beam, $e^{i\mathbf{k}\cdot\mathbf{r}}$. The second arises from the total wave function $\psi(\mathbf{r}')$ at some point $\mathbf{r}'$ being modified by multiplying by the potential $V(\mathbf{r}')$, and the resulting wave function developing as a spherical wave from $\mathbf{r}'$ to the point $\mathbf{r}$; the contributions from all possible points $\mathbf{r}'$ are added together linearly to give, with the first term, the total wave function at $\mathbf{r}$. A similar interpretation can be given to the corresponding integral equation for ψ_s which we can obtain from (6.28) by replacing ψ by $(e^{i\mathbf{k}\cdot\mathbf{r}}+\psi_s)$; this integral equation is

$$\psi_s(\mathbf{r}) = -\frac{2\pi m}{\hbar^2}\int\frac{e^{ik|\mathbf{r}-\mathbf{r}'|}}{|\mathbf{r}-\mathbf{r}'|}\,V(\mathbf{r}')e^{i\mathbf{k}\cdot\mathbf{r}'}\,d^3\mathbf{r}'$$

$$-\frac{2\pi m}{\hbar^2}\int\frac{e^{ik|\mathbf{r}-\mathbf{r}'|}}{|\mathbf{r}-\mathbf{r}'|}\,V(\mathbf{r}')\psi_s(\mathbf{r}')\,d^3r' \qquad (6.29)$$

The integrals on the right hand side of (6.29) may be interpreted as follows. The first integral comes from that part of the incident wave which is scattered once, i.e. multiplied by the potential at the point $\mathbf{r}'$, and then the result develops as a spherical wave to the point $\mathbf{r}$. The second integral comes from the scattering of the scattered wave itself by the potential at $\mathbf{r}'$ and then the result develops as a spherical wave to $\mathbf{r}$; again summation is performed over all such points $\mathbf{r}'$. This process is described graphically in Fig. 6.4. It is evidently important to try to solve the scattering integral equation (6.29), and we turn to that in the next section.

Fig. 6.4 The graphical description of the scattering equation (6.29); we denote by the left-hand side of the graphical equation the scattered wave $\psi_s(\mathbf{r})$ coming to the point $\mathbf{r}$, so by a heavy line coming in from the left to point $\mathbf{r}$. A dotted line coming in from the left to the point $\mathbf{r}$ denotes a plane wave $e^{i\mathbf{k}\cdot\mathbf{r}}$; on the right of the graphical equation it is scattered by the potential $V(\mathbf{r}')$, denoted by a cross at $\mathbf{r}'$ and propagates from $\mathbf{r}'$ to $\mathbf{r}$ by the propagator $e^{ik|\mathbf{r}-\mathbf{r}'|}/|\mathbf{r}-\mathbf{r}'|$, denoted by a thin line; the other term denotes the scattered wave itself coming in to $\mathbf{r}'$, being scattered there and propagating to $\mathbf{r}$. The contributions are summed over all $\mathbf{r}'$.

6.4 The Born Approximation

There are several ways of trying to solve the scattering integral equation (6.29). In this section we will consider a method of obtaining an approximate solution when the potential V is weak. By this we mean that there is a very small contribution arising from repeated scatterings of the particle beam by the potential, as contained in an implicit fashion in the second term on the right hand side of (6.29), or as described explicitly by Fig. 6.5. In other words, we need only keep the first term on the right-hand side of (6.29). This approximate solution of the equation is called the *Born approximation*; for the scattering amplitude $f(\theta)$ we see from our argument leading to (6.27)

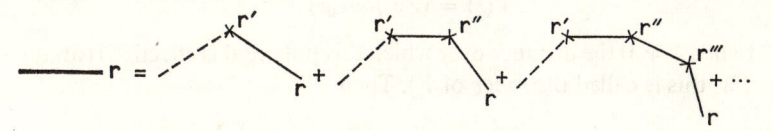

Fig. 6.5 A graphical representation of the iteration of the scattering equation (6.29), showing the first Born term as the first term in the right of the graphical equation, with following terms denoting higher Born terms.

that the Born approximation $f_B(\theta)$ to the scattering amplitude $f(\theta)$ is

$$f_B(\theta) = -(2\pi m/h^2) \int e^{i(\mathbf{k}-\mathbf{k}')\cdot\mathbf{r}'} V(\mathbf{r}') d^3\mathbf{r}'$$
$$= -(2\pi m/h^2) \tilde{V}(\mathbf{k}-\mathbf{k}') \qquad (6.30)$$

where $\tilde{V}(\mathbf{k})$ is the Fourier transform of the potential,

$$\tilde{V}(\mathbf{k}) = \int e^{i\mathbf{k}\cdot\mathbf{r}} V(\mathbf{r}) d^3\mathbf{r}$$

We may expect the Born approximation to work best at very high energies, since then the wave function $e^{i\mathbf{k}\cdot\mathbf{r}}$ oscillates very much, which may be shown to cause the value of $\psi_s(\mathbf{r}')$ in the last term on the right hand side of (6.29) to be heavily damped; this will occur even when the potential is relatively strong. In other words, at high energy the scattered wave is small, since there is cancellation between different parts of the spherical waves which combine to make it up in the manner we have already discussed.

As an example of this let us consider a spherically symmetrical potential $V(r)$ so that if we take polar co-ordinates with the z-axis along the direction $\mathbf{k}-\mathbf{k}'$, the value of $f_B(\theta)$ is

$$f_B(\theta) = -\frac{2\pi m}{h^2} \int_0^\pi d\theta' \int_0^\infty dr\, e^{iKr\cos\theta'} 2\pi r^2 V(r) \sin\theta'$$

where K, called the momentum transfer, has the value

$$K = |\mathbf{k}-\mathbf{k}'| = 2k \sin\tfrac{1}{2}\theta$$

as we see from Fig. 6.6. Then

$$f_B(\theta) = -\frac{2m}{\hbar^2} \int_0^\infty \frac{\sin Kr}{Kr} V(r) r^2 \, dr \tag{6.31}$$

If V is a screened Coulomb potential

$$V(r) = (Ze^2/4\pi\epsilon_0 r) e^{-\lambda r}$$

(when $\lambda \neq 0$ the distance over which this potential is effective is about $1/\lambda$; this is called the *range* of V). Then

$$f_B(\theta) = \frac{Ze^2}{8\pi\epsilon_0 mV^2} \operatorname{cosec}^2 \tfrac{1}{2}\theta \left[1 + \left(\frac{\lambda}{2k \sin \tfrac{1}{2}\theta}\right)^2\right]^{-1}$$

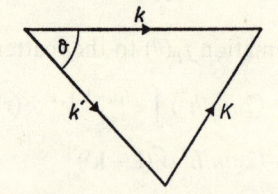

Fig. 6.6 The relation between the incoming momentum **k**, the outgoing momentum **k**′, the momentum transfer K and the scattering angle θ.

If $\lambda/2k \ll 1$, then

$$f_B(\theta) = \frac{Ze^2}{8\pi\epsilon_0 mV^2} \frac{1}{\sin^2 \tfrac{1}{2}\theta} \tag{6.32}$$

Then (6.32) gives the classical differential cross section (6.6) for a Coulomb field, on forming $|f_B(\theta)|^2$.

If the Born approximation is not satisfactory, it is possible to go to the *second* Born approximation, which is obtained not by neglecting the second term in the right hand side of (6.29), but by taking for $\psi_s(\mathbf{r}')$ in that term the first Born approximation value, equal to the first term on the right hand side of (6.29). Then the second Born approximation $f_B^{(2)}(\theta)$ to the scattering amplitude is

$$f_B^{(2)}(\theta) = \frac{-2\pi m}{\hbar^2} \tilde{V}(\mathbf{k}-\mathbf{k}')$$

$$+ \left(\frac{2\pi m}{\hbar^2}\right)^2 \int e^{-i\mathbf{k}'\cdot\mathbf{r}'} V(\mathbf{r}') \frac{e^{ik|\mathbf{r}'-\mathbf{r}''|}}{|\mathbf{r}'-\mathbf{r}''|} e^{i\mathbf{k}\cdot\mathbf{r}''} V(\mathbf{r}'') d^3\mathbf{r}' \, d^3\mathbf{r}''$$

It is possible to go to the third Born approximation in an evident

SCATTERING THEORY

fashion; the higher Born approximation may be obtained by iterating (6.29), regarding V as small; this is shown graphically in Fig. 6.5.

6.5 Phase Shifts

The phase shift analysis is a useful approach at low energies, where the Born approximation has more chance of breaking down. The phase shifts are introduced by considering the total wave function $\psi(\mathbf{r})$ which describes a scattering process, not as the sum of an incident plane wave and an outgoing spherical wave, but as the sum of incoming and outgoing spherical waves. To achieve this decomposition, we treat the asymptotic form (6.8) for the wave function as valid for large distances $|\mathbf{r}|$ from the scattering centre. We know that spherical wave solutions of the Schrödinger equation will have the separable form $[u(r)/r]Y_l^m(\theta,\phi)$, where $u(r)$ satisfies the one dimensional radial Schrödinger equation

$$\frac{d^2 u}{dr^2} + \left\{k^2 - \frac{l(l+1)}{r^2}\right\}u = 0 \qquad (6.33)$$

(where we neglect the potential in (6.33)). For a very large r we take the radial wave function of the scattered part of ψ to be just an outgoing wave term e^{ikr}; since the scattered wave function does not depend on ϕ we take $m = 0$, so have the term $(e^{ikr}/r)P_l(\cos\theta)$. Since any linear combination of such terms also satisfies the free Schrödinger equation, we may take the scattered part of the wave function to be $\sum_l a_l P_l(\cos\theta)(e^{ikr}/r)$, where a_l only depends on l and k. Then we may take for the scattering amplitude $f(\theta)$ the form

$$f(\theta) = \sum a_l P_l(\cos\theta) \qquad (6.34)$$

In order to obtain some condition on the coefficients a_l, we also attempt to expand the incoming plane wave part $e^{i\mathbf{k}\cdot\mathbf{r}}$ in (6.8) in a similar fashion. We take the z-axis along the direction of $\mathbf{k}$, so that we have to find the expansion of $e^{ikr\cos\theta}$ in Legendre polynomials:

$$e^{ikr\cos\theta} = \sum b_l P_l(\cos\theta) \qquad (6.35)$$

For large r, $b_l = i^l(2l+1)\sin(kr - \tfrac{1}{2}l\pi)/kr$; this can be seen if we use that $\int_{-1}^{+1} P_l(x)P_m(x)\,dx = 2\delta_{lm}/(2l+1)$, so that multiplying both sides of (6.35) by $P_m(\cos\theta)$ and integrating over θ:

$$2b_m/(2m+1) = \int_{-1}^{+1} P_m(x)e^{ikrx}\,dx \tag{6.36}$$

We integrate the right-hand side of (6.36) by parts to give

$$\frac{2i^l}{kr}\sin(kr - \tfrac{1}{2}l\pi) - \frac{1}{kri}\int_{-1}^{+1} e^{ikr}P'_l(x)\,dx \tag{6.37}$$

and the second term of (6.37) is much smaller than the first for large r (it decreases by one power of r faster than the first term), so that putting together (6.36) and the first term in (6.37) gives the value of b_l we stated above. Then for large r the total wave function has the form

$$\psi(r) = \frac{1}{kr}\sum_l P_l(\cos\theta)\left\{i^l(2l+1)\sin(kr-\tfrac{1}{2}l\pi) + ka_l e^{ikr}\right\}$$

$$= \frac{1}{kr}\sum_l P_l(\cos\theta)\left[e^{ikr}\left\{ka_l + \frac{(2l+1)}{2i}\right\} - e^{-ikr}(2l+1)\frac{i^{2l}}{2i}\right] \tag{6.38}$$

The last line of (6.38) gives the expansion of ψ as a sum of an incoming spherical wave (proportional to e^{-ikr}/r) and an outgoing spherical wave (proportional to e^{ikr}/r). Since there is no loss of probability, the incoming flow of probability carried by the incoming spherical wave must equal the outgoing flow carried by the outgoing spherical wave; this must be true for each separate term in the summation over l in (6.38) (because these separate terms are independent, since the P_l-functions are, and carry differing values of total angular momentum l). Thus

$$\left|ka_l + \frac{2l+1}{2i}\right| = \frac{(2l+1)}{2}$$

or

$$ka_l + \left(\frac{2l+1}{2i}\right) = e^{2i\delta_l}\left(\frac{2l+1}{2i}\right) \tag{6.39}$$

where δ_l is some real constant. We choose δ_l so that it vanishes when a_l is zero, as, for example, when the potential V vanishes. Then

$$a_l = \frac{(2l+1)}{2ik}(e^{2i\delta_l} - 1)$$

and
$$f(\theta) = \frac{1}{k} \sum_{l} (2l+1) P_l(\cos\theta) e^{i\delta_l} \sin\delta_l \qquad (6.40)$$

The expression (6.40) is called the *phase shift decomposition* of the scattering amplitude, where δ_l is called the *phase shift* (for the lth partial wave or angular momentum state). We have thus given a parametrization of the scattering amplitude in terms of the phase shift. Each of these is a real function of energy; the value for a given potential V of δ_l for all l will give the scattering amplitude by (6.40). Thus we have reduced our problem to calculating the phase shifts.

The set of phase shifts are an infinite set of real numbers; it is to be expected, however, that only a finite number of these phase shifts will be important for scattering at a given energy, and at very low energy only one phase shift will be important, that for $l = 0$. This is because if the potential has a finite range r_0 i.e. a distance beyond which it is negligible, we expect that only angular momenta up to $\hbar k r_0$ will arise for scattered particles, so that l takes values up to $l_{\max}$ with

$$\hbar l_{\max} = \hbar k r_0 \qquad (6.41)$$

Then as $k \to 0$, so that the energy of the incoming particle is very small, the value of $l_{\max}$ in (6.41) will be zero; there will only be *s*-wave scattering in this case.

We may write down the differential cross section $d\sigma/d\Omega$ and the total cross section σ in terms of the partial wave expansion (6.40):

$$\frac{d\sigma}{d\Omega} = \frac{1}{k^2} |\sum_{l} (2l+1) P_l(\cos\theta) e^{i\delta_l} \sin\delta_l|^2 \qquad (6.42)$$

$$\sigma = \frac{1}{k^2} \int d\Omega |\sum_{l} (2l+1) P_l(\cos\theta) e^{i\delta_l} \sin\delta_l|^2$$

$$= \frac{1}{k^2} \int d\Omega \sum_{l,m} (2l+1)(2m+1) P_l(\cos\theta) P_m(\cos\theta) e^{-i\delta_l} \sin\delta_l e^{i\delta_m} \sin\delta_m \qquad (6.43)$$

If we use the property that $\int_{-1}^{+1} P_l(x) P_m(x)\, dx = 2\delta_{lm}/(2l+1)$ we see that the total cross section of (6.43) is

$$\sigma = \frac{4\pi}{k^2} \sum_{l} (2l+1) \sin^2 \delta_l \qquad (6.44)$$

At very low energy when only the s-wave, $l = 0$, contributes in the summation over l in (6.42) and (6.44) we have

$$\frac{d\sigma}{d\Omega} = \frac{\sin^2 \delta_0}{k^2}, \qquad \sigma = \frac{4\pi}{k^2} \sin^2 \delta_0 \qquad (6.45)$$

Evidently the scattering in the low energy region is isotropic, since $d\sigma/d\Omega$ is independent of θ. This is to be expected, since for low energy the wave length λ of the incident particle becomes very large ($\lambda = 1/k$). The scatterer can only be sensible to the direction of the incident particle if the phase difference of the incident wave over the size of the scattering region is not negligible, otherwise the scattering potential behaves like a point scatterer. But it is precisely such behaviour which will arise if the wave length of the particle is much larger than the effective size of the scattering region; such a 'point' scatterer can only give isotropic scattering, the incident beam having essentially one phase over the whole of the scattering region.

There is a further important property of the very low energy scattering. We saw that the scattered wave is produced by the effect of the potential on the incident wave; this effect arises in the Schrödinger equation through the term $(E - V(r))$, where the centrifugal term does not contribute for $l = 0$. If the energy is very small, then it is legitimate to neglect E in comparison with $V(r)$, so that the scattered wave becomes independent of E, or of k, as $k \to 0$. This means that the scattering amplitude $f(\theta)$ approaches a constant value which we denote by $-a$, where a is called the *scattering length*:

$$\lim_{k \to 0} f(\theta) = -a \qquad (6.46)$$

so

$$\lim_{k \to 0} \delta_0 = -ak \qquad \text{(to within multiples of } \pi\text{)}$$

and

$$\lim_{k \to 0} \sigma = 4\pi a^2$$

6.6 The Square Well

In order to understand the relation between the wave function and low energy scattering, in particular the way bound states and resonances affect the phase shifts, we will consider an exactly soluble model

SCATTERING THEORY

which we discussed before in connection with bound states—the square well potential, (see §3.4). This potential is very useful for illustrating the scattering of sub-nuclear particles by one another, so has more than pedagogical interest.

The one dimensional radial Schrödinger equation for the three-dimensional square well potential

$$V(r) = V_0 \quad (r \leq r_0) \\ = 0 \quad (r > r_0) \qquad (6.47)$$

for the s-wave radial part $u_0(r)$ of the wave function is

$$\frac{d^2 u_0}{dr^2} + \frac{2m}{\hbar^2}(E - V_0)u_0 = 0 \quad (r \leq r_0) \qquad (6.48)$$

$$\frac{d^2 u_0}{dr^2} + \frac{2m}{\hbar^2} E u_0 = 0 \quad (r > r_0) \qquad (6.49)$$

The boundary conditions on the wave function $u_0(r)$ are that

(i) $u_0(0) = 0$ \hfill (6.50)

(ii) $u_0, du_0/ds$ are continuous at $r = r_0$. \hfill (6.51)

The condition (6.50) arises because if $u_0(0) \neq 0$, then the radial dependence of the total wave function would be like $1/r$ as $r \to 0$; we know that the function $(-1/4\pi r)$ is a solution of $\nabla^2(-1/4\pi r) = \delta^3(\mathbf{r})$, and this δ-function could not be compensated by any term in the potential (6.47), so that the complete Schrödinger equation would not be satisfied. Then the solution of (6.48) will be

$$u_0 = A \sin \kappa r, \quad \kappa = \sqrt{\{2m(E - V_0)/\hbar^2\}} \quad (r \leq r_0) \qquad (6.52)$$

and of (6.49) will be some linear combination of $\sin kr$ and $\cos kr$, where $k = \sqrt{2mE/\hbar^2}$ (and we remember that the energy E is positive, so k is real; we may, however, have κ imaginary when V_0 is positive and larger than E). We take

$$u_0 = B \sin(kr + \delta_0) \quad (r > r_0) \qquad (6.53)$$

where δ_0 is a constant we will identify with the s-wave phase shift shortly. We now apply the boundary condition (6.51) to the solutions (6.52) and (6.53) to give

$$B\sin(kr_0+\delta_0) = A\sin\kappa r_0 \tag{6.54}$$

$$B\cos(kr_0+\delta_0) = A\kappa\cos\kappa r_0 \tag{6.55}$$

Dividing the left-hand side of (6.55) by that of (6.54), and equating to the ratio of the right-hand side, we obtain the condition

$$\kappa\cot\kappa r_0 = k\cot(kr_0+\delta_0) \tag{6.56}$$

Now from (6.38) and (6.39) we know that the asymptotic form of $\psi(\mathbf{r})$, arising purely from the s-wave part, is

$$\psi(\mathbf{r}) \to (e^{i\delta_0}/kr)\sin(kr+\delta_0)$$

so that in (6.53) we see that the constant δ_0 we introduced there is correctly identified with the s-wave phase shift.

Let us now consider the wave function for low energy scattering; we show in Fig. 6.7 the radial function $u_0(r)$ startingout with $\sin\kappa r$ from $r = 0$ to $r = r_0$, or $\sinh\kappa r$ for a repulsive potential when $E < V_0$, and $\kappa = \sqrt{\{2m(V_0-E)\}/\hbar}$, then joining on smoothly to $\{\sin\kappa r_0/\sin(kr_0+\delta_0)\}\sin(kr+\delta_0)$ for $r > r_0$, or $\{\sinh\kappa r_0/\sin(kr_0+\delta_0)\}\sin(kr+\delta_0)$ for $E < V_0$ in the repulsive case. We see that for a repulsive potential, with $\kappa < k$, the wave function is repelled from the interior region, as is expected physically. Since the function $\sin(kr+\delta_0)$ is approximately $k(r-a)$ for small r, we expect, for either sign of V_0, that a linear continuation of the exterior wave function (6.55) to $r < r_0$ will have a zero at $r = a$. In the repulsive potential case (ii) of Fig. 6.7, we see from the figure that a is positive, for $-\delta_0/k$ is the value of the shift of the node from its original value π/k, and since this shift is positive then δ_0 is negative (so a is positive). In the case (iii) of an infinitely repulsive potential at r_0, the whole free wave function $\sin kr$ is shifted to the right by a distance r_0; thus

$$\lim_{k\to 0}(-\delta_0/k) = r_0, \qquad a = -r_0$$

$$\lim_{k\to 0}\sigma = 4\pi r_0^2$$

This is four times the geometrical cross section πr_0^2 (which is the area presented to the particle by the scatterer).

In the case of an attractive square well with V_0 negative, as shown in cases (iv), to (vi) of Fig. 6.7, we see that the wave function is pulled into the well, with δ_0/k positive, since now $\kappa > k$. As V_0 becomes more negative we see that δ_0 will increase; just before its value becomes $\pi/2$ (actually equal to $\pi/2 - kr_0$), the radial function $u_0(r)$ has zero slope at r_0, and appears as in case (v) of Fig. 6.7. In this case the total cross section is approximately $4\pi/k^2$, which is infinite for $k = 0$; this behaviour of the cross section is seen in the anomalously large cross section for the scattering of very slow neutrons by protons. As V_0 becomes more negative the slope of the interior wave function at r_0 becomes negative. We may make the energy slightly negative without changing this interior wave function essentially, and then join the interior wave function to a decreasing exponential for $r > r_0$. This represents a bound state wave function, the bound state having very little binding energy. As we further decrease V_0, the binding energy of this bound state increases. The phase shift δ_0 will also increase; near zero energy we will reach a value of V_0 for which the slope of the interior wave function at r_0 is again zero, as shown in case (vi). The value of δ_0 for this to happen is $(3\pi/2) - kr_0$; again, for more negative V_0 we may join up an exponentially decreasing wave function of slightly negative energy to the interior wave function so that the potential well can now sustain another bound state of small energy. Continuing in this fashion we see that each time the phase shift $\delta_0(0)$ at zero energy increases through $(n+\frac{1}{2})\pi$, due to V_0 becoming more negative, an extra bound state of zero energy may be sustained by the square well; the total number of bound states sustained by the potential when $(n-\frac{1}{2})\pi < \delta_0(0) < (n+\frac{1}{2})\pi$ is equal to n. We may indeed write the solution to (6.56) as

$$\delta_0 = n\pi - kr_0 + \tan^{-1}\{(k/\kappa)\tan \kappa r_0\} \quad (6.57)$$

where n is the number of bound states and the inverse tangent is an angle in the range of $(-\pi/2, +\pi/2)$.

6.7 Bound States and Resonances

The discussion given above, for the dependence of the s-wave phase shift on the strength of the scattering potential in the particular case of a square well potential, has not depended crucially on the exact solution for the wave function; we expect the same general properties of the solution to be valid for a general short range potential. We have already seen how bound states may appear; we will now discuss this

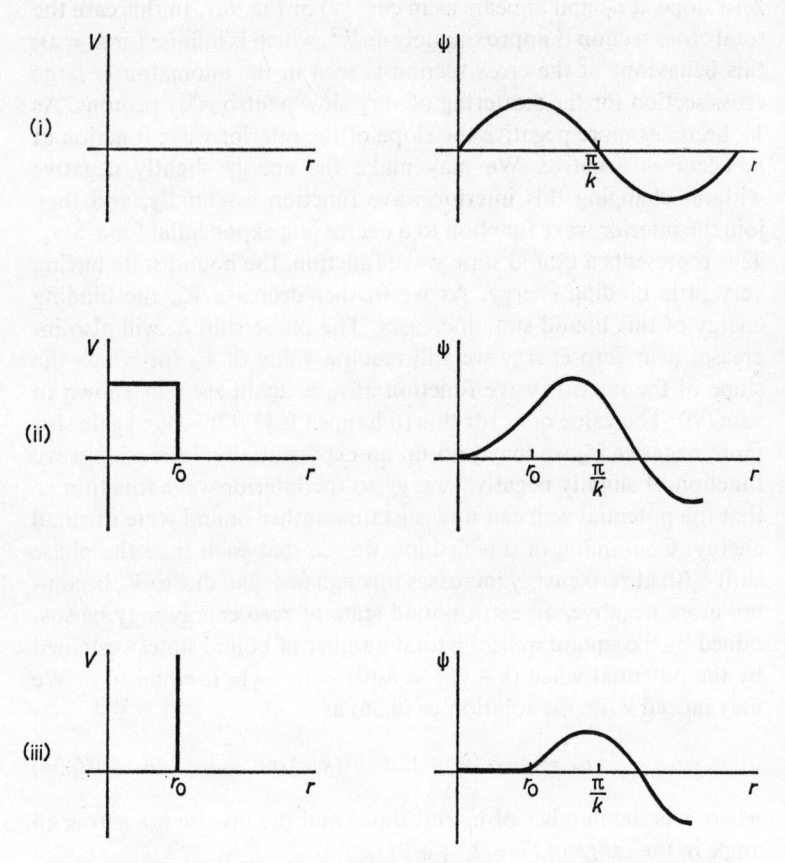

Fig. 6.7 A graph of the suitably normalized wave function describing scattering by a square well for various cases; the wave function inside the well ($r < r_0$) is taken to be $\sin \kappa r$ ($\sinh \kappa r$ for a repulsive well with $E < V_0$), joining for $r > r_0$ to $[\sin \kappa r_0/\sin(kr_0 + \delta_0)] \sin(kr + \delta_0)$ (or $[\sinh \kappa r_0/\sin(kr_0 + \delta_0)] \sin(kr + \delta_0)$ for the repulsive case).
(i) Vanishing square well; wave function has first node at π/k.
(ii) Repulsive potential with $E < V_0$ so that $\kappa < k$ and the wave function is pushed out of the potential region; the first node is moved to the right of π/k.
(iii) An infinitely repulsive potential at r_0, having zero wave function inside the potential region.

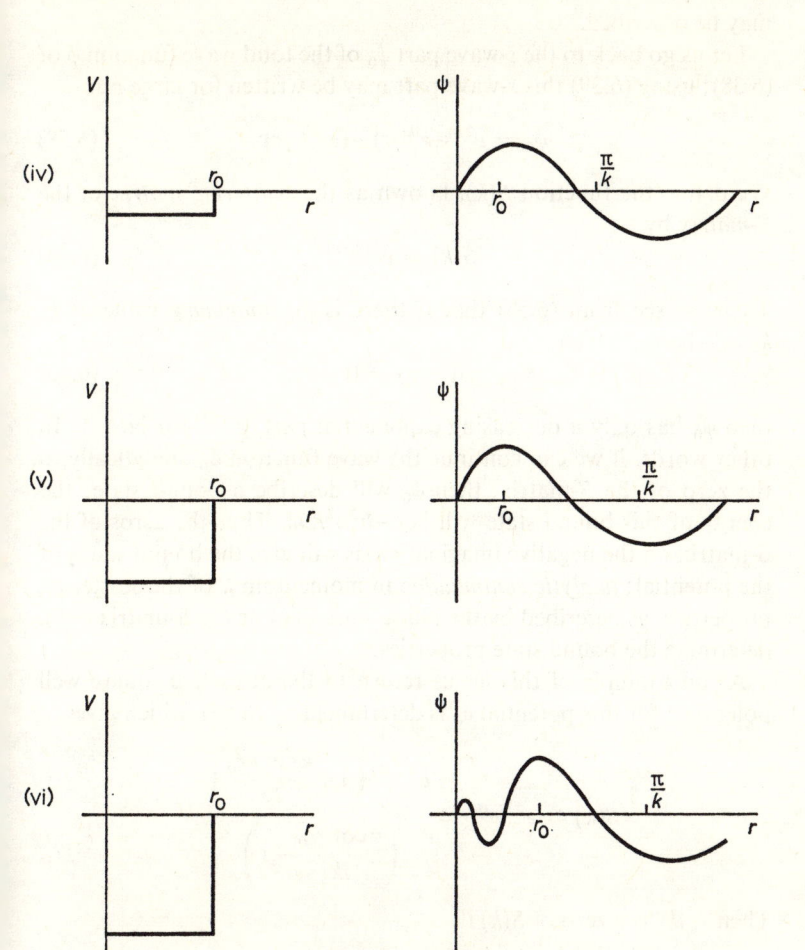

(iv) An attractive potential, the wave function being pulled into the potential region with the first node moved to the left of π/k.
(v) A more attractive potential than (iv), strong enough to give zero slope to the wave function at $r = r_0$; this just allows a zero energy bound state to be trapped by the potential well.
(vi) An even more attractive potential than (v) with slope again zero at $r = r_0$, now trapping a second zero energy bound state.

in more detail, and indicate how decaying states, called resonances, may be described.

Let us go back to the *s*-wave part ψ_0 of the total wave function ψ of (6.38); using (6.39) this *s*-wave part may be written for large r as

$$\psi_0 = e^{2i\delta_0}(e^{ikr}/r) - (e^{-ikr}/r) \tag{6.58}$$

We define the function $S(k)$, known as the *scattering matrix* or the *S-matrix* by

$$S(k) = e^{2i\delta_0(k)} \tag{6.59}$$

Then we see from (6.58) that if there is an *imaginary* value of k, $k = -i\kappa$, for which

$$S(-i\kappa) = 0 \tag{6.60}$$

then ψ_0 has only a decreasing exponential part, $e^{-\kappa r}$, for large r. In other words, if we can continue the wave function ψ_0 analytically to the zero of the *S*-matrix, then ψ_0 will describe a bound state; the energy of this bound state will be $-\hbar^2\kappa^2/2m$. Thus the zeros of the *S*-matrix on the negative imaginary axis will give the bound states of the potential; *analytic continuation* in momentum k of the scattering properties, as described by the phase shift $\delta_0(k)$ or the *S*-matrix $S(k)$, determine the bound state properties.

As an example of this let us return to the attractive square well potential; for this potential δ_0 is determined by (6.56), which gives

$$S(k) = e^{2i\delta_0(k)} = \frac{e^{-2ikr_0}\left(1 + \dfrac{\kappa\cot\kappa r_0}{ik}\right)}{\left(\dfrac{\kappa\cot\kappa r_0}{ik} - 1\right)}$$

Then $-ik'$ is a zero of $S(k)$ if

$$(k'/\kappa)\tan\kappa r_0 + 1 = 0 \tag{6.61}$$

This may be verified as the correct *s*-wave bound-state condition for a three-dimensional potential well by applying the usual continuity conditions for the radial wave function $u(r)$ at $r = r_0$, i.e. u and du/dr are continuous at $r = r_0$, where the radial Schrödinger equation

$$\frac{d^2u}{dr^2} + \frac{2m}{\hbar^2}(E - V_0)u = 0$$

SCATTERING THEORY

has solutions

$$u = A \sin \kappa r \quad (r < r_0)$$
$$u = e^{-k'r} \quad (r \geq r_0) \quad (6.62)$$

For then we require

$$A \sin \kappa r_0 = e^{-k'r_0}$$
$$\kappa A \cos \kappa r_0 = -k' e^{-k'r_0} \quad (6.63)$$

which has a non-zero value for A if (6.61) is fulfilled.

Let us consider further properties of $S(k)$ in the complex k-plane in the effective range approximation which is valid for low energies. This approximation is achieved by expanding the function $k \cot \delta_0$ in powers of k. The function is even in powers of k; this is true for the square well, as can be seen by inspection of (6.56), and if we take it to hold more generally, we may write for small k:

$$k \cot \delta_0 = -(1/a) + \tfrac{1}{2} R k^2 \quad (6.64)$$

The quantity a is the scattering length which we introduced earlier; R is called the effective range, and may be shown to be positive for an attractive potential. For the square well it may be shown to be

$$R = r_0 \left\{ 1 - \frac{1}{(Kr_0)^2} \left(\frac{r_0}{a} \right) - \frac{1}{3} \left(\frac{r_0}{a} \right)^2 \right\}$$

where $K^2 = -2mV_0/\hbar$. From (6.64) we have that

$$\sigma = \frac{4\pi \sin^2 \delta_0}{k^2} = \frac{4\pi}{k^2(1 + \cot^2 \delta_0)} = \frac{4\pi}{k^2 + (-1/a + \tfrac{1}{2} R k^2)^2} \quad (6.65)$$

whilst the S-matrix is

$$S(k) = \frac{1/a - \tfrac{1}{2} R k^2 - ik}{1/a - \tfrac{1}{2} R k^2 + ik} \quad (6.66)$$

We see that (6.66) satisfies $|S(k)| = 1$ for real k, as it should. If $R = 0$ then we see that $S(k)$ has a zero at $k = i/a$; this will be interpreted as a bound state, as we described earlier, if a is negative (an attractive potential). In this case $S(k)$ has a pole at $k = i/a$, so we may alternatively describe bound states by poles on the positive imaginary axis. This is true quite generally, since $|S(k)| = 1$ so for complex k, $S(k) = S^*(k^*)^{-1}$ and a zero of $S(k)$ will give a pole of S^* (so S) at the complex conjugate position. When R is non-zero, $S(k)$ will have two zeros at the values of k equal to

$$\frac{1}{R}\left\{-i \pm \sqrt{2\left(\frac{R}{a}-1\right)}\right\}$$

These zeros may be interpreted in general as describing pseudo bound states, or resonances, as they are called; these states give important contributions to the scattering amplitude, though they are not stable with an infinite lifetime, as are bound states. We saw that a bound state of very small binding energy gives a phase shift δ_0 at zero energy of $\pi/2$, so giving an infinite cross section there. This large cross section arises since in this case a is very large; the resulting s-wave scattering amplitude is (neglecting Rk^2 compared with k)

$$f_0(\theta) = (1/2ik)(S(k)-1) = 1(1/a+ik) \tag{6.67}$$

This has the bound state pole at $k = -i/a$, and it is this which gives the increase in the cross section at very low energy. Let us consider the nature of the contribution from a possible pole near the *real* axis; regarding this pole as in the energy variable E (which is equivalent to k, since $E = \hbar^2 k^2/2m$, but is more convenient to use, especially in discussing the time dependence of wave functions) $f_0(\theta)$ would then have the form

$$f_0(\theta) = \Gamma/(E-E_r+i\Gamma) \tag{6.68}$$

where we take $E_r - i\Gamma$ to be the position of the pole, in the complex energy plane. The corresponding cross section

$$\sigma = 4\pi|f_0(\theta)|^2 = \frac{4\pi\Gamma^2}{(E-E_r)^2+\Gamma^2}$$

has a peak at $E = E_r$; if Γ is not too large this peak is very noticeable in the energy dependence of both the differential and total cross sections. We supposed that such a pole described a semi-bound particle of energy $(E_r - i\Gamma)$ in a similar fashion to a bound state being described by a pole, though now on the negative energy axis. The time dependence of the wave function for such a particle is then $e^{-iE_r t/\hbar - \Gamma t/\hbar}$, so decreases exponentially, with lifetime of order $\hbar/\Gamma$. We see that the resonance poles must always be at points $E_r - i\Gamma$, with Γ positive, since otherwise the time dependent wave function would increase exponentially in time, which is impossible. If the resonance pole is close to the real energy axis (with Γ small) it will describe a long-lived particle and show up as a sharp peak in the energy variation of the differential and total cross sections; if it is far away from the real axis, with Γ large, it will describe a short lived particle, and only give a broad peak in the energy dependence of cross sections. The two

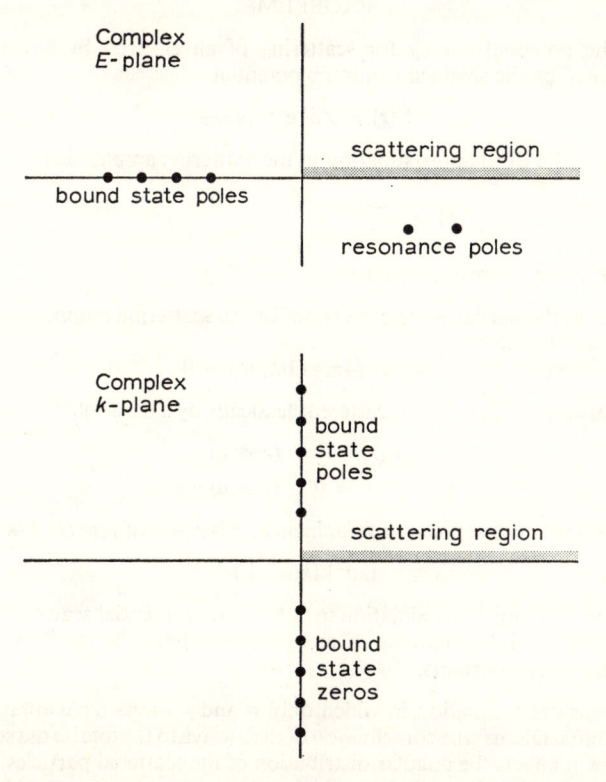

Fig. 6.8 Positions of bound states and resonance poles in the complex energy plane which arise in the S-matrix, when regarded as a function of the complex energy E.
Also positions of bound state poles and zeroes in the complex momentum plane for the S-matrix.

extreme ranges of such behaviour do actually arise in the sub-nuclear particles. For example the host of resonances closely similar to protons and neutrons have life times of order 10^{-22} seconds, with $\Gamma \sim 100$ MeV; the μ-meson is a long lived particle decaying in a fashion similar to radioactive decays, and lives for 10^{-10} seconds. The general positioning of bound states and resonances in the energy plane and in the complex k-plane is shown in Fig. 6.8.

PROBLEMS

6.1 The potential energy for scattering of an electron by an atom is represented by the shielded Coulomb potential

$$V(\mathbf{r}) = Ze^2 \, e^{-r/a}/4\pi\epsilon_0 r$$

Show that the Born approximation to the scattering amplitude is

$$f(\theta) = \frac{2mZe^2}{4\pi\epsilon_0 \hbar^2 K^2}\left[1 - \frac{1}{1+(Ka)^2}\right]$$

where K is the momentum transfer in scattering.

6.2 From the partial wave expansion for the scattering amplitude, prove the 'optical theorem':

$$\sigma_{\text{tot}} = (4\pi/k)\,\text{Im} f(\theta) = 0$$

6.3 Particles of mass m are scattered classically by a hard sphere potential:

$$V(r) = \infty \quad (r \leqslant a)$$
$$= 0 \quad (r > a)$$

Show that the p-wave phase shift for incident particles with energy $\hbar^2 k^2/2m$ is

$$\tan^{-1}(ka) - ka$$

6.4 Use the Born approximation to obtain the differential scattering cross section for a spinless particle of energy E by a potential Ae^{-br}, where A and b are given constants.

6.5 Consider a situation in which only s- and p-waves have appreciable phase shifts. Discuss the contribution of the p-wave to the total cross section, and how it affects the angular distribution of the scattered particles. What sort of measurement should be made to obtain an accurate value of δ_0 or δ_1. How might a small d-wave phase shift be detected?

FURTHER READING

1. GOLDBERGER, M. and WATSON, K., *Collision Theory*, John Wiley, 1964; a very complete account of the theory of scattering.
2. ZIENAU, S., *Scattering Theory*, Allen and Unwin, 1970/1; to appear in this series.

CHAPTER 7

Conclusion

7.1 Introduction

The contents of this book are regarded as forming the minimum basis of quantum mechanics for a student learning modern physics. However, certain important areas of quantum mechanics have had to be left out or treated rather cursorily, as is to be expected in an introductory book of this size. It is hoped that the bibliographies given at the end of each chapter will enable the student to read further into areas only cursorily treated. Even with this brief treatment of certain areas, the main thesis of the book, that quantum mechanics is a successful description of the behaviour of matter down to distances of about 10^{-10} metres, has been presented with reasonable detail. In particular we have had discussions of the quantum mechanical aspects of the electron theory of solids (in §3.7), of symmetries (in Chapter 4) of the electronic binding in atoms and molecules (in Chapter 5), and of scattering theory (in Chapter 6).

The most important area of quantum mechanics which we have had to leave out completely is the quantum mechanical aspects of the motion of fast particles, as described by special relativity. The very beautiful theory of Dirac, describing the relativistic motion of the electron and at the same time describing its intrinsic spin of one half, is an extremely important part of this area. The mathematical form of this theory, as expressed by the Dirac equation, enables relativistic effects to be satisfactorily taken account of for the motion of electrons in atoms; at the same time the Dirac equation predicts the existence of the *positron*, which is a particle identical to the electron but with an opposite electrical charge. The discovery of the positron in 1932 and of the anti-proton (the corresponding counterpart of the proton) in 1955 has been the outstanding verification of Dirac's theory. Since this theory is basic to an understanding of the more recent developments of the structure of matter at even smaller distances, of about 10^{-15} metres or less, which stem from the discovery of the host of excited states of protons, and the mesons which mediate the nuclear forces, holding the constituents of nuclei together, it was left out of this book only with deep reluctance. The brief discussion of electron spin and the exclusion principle, given in §5.1, and the discussion of the rotation group in §4.6, may go a little way towards making up for this omission, but cannot go far enough. Further reading on the Dirac theory is given at the end of this chapter, and the following remarks will help towards an understanding of the theory.

7.2 The Dirac Theory of the Electron

The main idea of the theory of Dirac is that there is a *representation* of the group of rotations of space, which we discussed in §4.6, by means of 2×2 matrices. This representation is intermediate between the scalar representation by a 1×1 matrix and the vector representation by 3×3 matrices (acting on 3-component vectors). This representation by means of 2×2 matrices associates with the angular momentum operators L_x, L_y, L_z the three matrices

$$\sigma_x = \frac{\hbar}{2}\begin{pmatrix} 0 & 1 \\ 1 & 0 \end{pmatrix}, \quad \sigma_y = \frac{\hbar}{2}\begin{pmatrix} 0 & i \\ -i & 0 \end{pmatrix}, \quad \sigma_z = \frac{\hbar}{2}\begin{pmatrix} 1 & 0 \\ 0 & -1 \end{pmatrix}$$

which satisfy the basic commutation rules of angular momentum operators (4.6), so may be regarded as angular momentum operators in their own right. The 2×2 matrix describing a rotation through an angle θ about the axis with unit vector **n** is then $\exp[(i/2)\theta\sigma \cdot \mathbf{n}]$, in a similar fashion to (4.77). These 2×2 matrices act on 2-component vectors, called (Pauli) *spinors*. It is these spinors which are used to describe the states of the spinning electron, one component corresponding to an electron with spin up, the other to an electron with spin down. When we try to describe an electron with a given *parity* it turns out that we have to double up the components of the spinor, so obtaining a (Dirac) Spinor ψ with four components, denoted by

$\psi_\alpha(\alpha = 1, 2, 3, 4)$. This spinor has to satisfy a linear differential equation if space and time are treated on the same footing:

$$\{\sum_{\mu=1}^{4} (\gamma_\mu \partial/\partial x_\mu) + imc/\hbar\}\psi = 0 \tag{7.1}$$

where $\gamma_1, \gamma_2, \gamma_3, \gamma_4$ are four 4×4 matrices which satisfy the anticommutation relations

$$(\gamma_\mu \gamma_\nu + \gamma_\nu \gamma_\mu)_{\alpha\beta} = 2\delta_{\mu\nu}\delta_{\alpha\beta} \tag{7.2}$$

and $x_4 = t$, the time. The Dirac spinor contains four independent components, so describes four independent states of the electron. We expect that these differ by spin up or spin down for the electron; the property which gives the other label to these four states is the sign of the energy—a positive energy state (with spin up or spin down) describing the usual electron, the negative energy states (again with spin up or down) now being used to describe the positron, though with positive energy. The positron can be shown to have opposite charge to the electron, and is called its *antiparticle*.

Besides this prediction of antiparticles, which as already remarked has been amply verified, the Dirac equation (7.1) may be extended to describe an electron in motion in an atom, or even a set of electrons in motion in an atom. In this manner extra terms may be shown to arise in the Hamiltonian for the system of electrons (where this Hamiltonian can *always* be introduced to describe the time development of the system of one or many electrons). These extra terms give corrections which explain, among other things, the more detailed structure of the periodic table, as we briefly mentioned at the end of §5.6.

7.3 Whither Quantum Mechanics?

In order to justify the extra time spent on the Dirac equation in this conclusion, and in order to emphasize the nature of the areas of modern research, it might be useful to distinguish the two lines of development in modern physics which use quantum mechanics, and which are of very different nature. One line of development is to solve more and more complicated problems about solids, liquids, and gases, using the known Hamiltonian describing the Coulomb interactions between the nuclei and the electrons in the various atoms making up the material. These problems can be very complicated, and their solutions can give a very important understanding of the nature of forces and reactions in various types of matter. This line of development may be regarded as building bigger and more beautiful houses and buildings with a *known* or *given* set of building blocks (unlimited in their number but limited in kind).

The other line of development, following the analogy just given, investigates the problem of understanding better how the building blocks themselves are made and so of getting better building blocks. In other words, the inner structure of the nuclei and the electrons are investigated, as is the Coulomb potential between these particles. Investigations are made at shorter and shorter distances and so at higher and higher energies (the de Broglie relation (2.2) shows that the wave length decreases with increasing energy, and the shorter the wave length, the more details of structure which can be picked out by a beam directed onto the object being investigated). This line of development leads always into the unknown; as described in §2.1, it has led to the change from quantum mechanics to quantum field theory. The latter theory is based on the formalism we have set up in this book, but allows for the annihilation or creation of particles. This process, which is seen in nature, allows potentials to be determined from the quantum field theory except for one or two constants, instead of being given from outside, as in the discussion in this book. The quantum field theory has had numerous successes, though it presents enormous challenges which still have hardly been faced up to. The author regards it as the most important line of development for the future of physics; it is here that the most fundamental changes occur in our understanding of the physical world.

FURTHER READING

1. Rose, M., *Relativistic Electron Theory*, John Wiley, 1961.
2. Dirac, P. A. M., *Quantum Mechanics*, Clarendon Press, 1957.

APPENDIX 1

Angular Momentum

We wish to show that if **L** is given by (4.1) then L_z can only have the eigenvalues $(0, \hbar, 2\hbar, \cdots)$. To do this we write L_z as

$$L_z = xp_y - yp_x = C_+ C_- - (A_+ A_- + B_+ B_-) \tag{A.1}$$

where

$$A_\pm = \frac{1}{\sqrt{2}}(p_x \mp ix), \quad B_\pm = \frac{1}{\sqrt{2}}(p_y \mp iy), \quad C_\pm = B_+ \pm iA_+.$$

Then the following commutation relations may be established by means of the basic commutation relations of **r** and **p** (2.45):

$$[A_+, A_-]_- = 1 \tag{A.2}$$

$$[B_+, B_-]_- = 1 \tag{A.3}$$

$$[C_+, C_-]_- = 2 \tag{A.4}$$

$$[A_- A_+, B_- B_+]_- = 0 \tag{A.5}$$

$$[C_- C_+, A_- A_+ + B_- B_+]_- = 0 \tag{A.6}$$

Then since from (A.6), the operator $C_- C_+$ commutes with $A_- A_+ + B_- B_+$, the eigenvalues of L_z are the differences of the eigenvalues of $C_- C_+$ and $A_- A_+ + B_- B_+$; from (A.5) the eigenvalues of this latter operator are the sum of those of $A_- A_+$ and $B_- B_+$. So we are left with finding the eigenvalues of the three operators $A_- A_+$, $B_- B_+$ and $C_- C_+$ where (A.2), (A.3) and (A.4) are satisfied. From the similarity of these three cases we need only consider $A_- A_+$, say; $B_- B_+$ will have identical eigenvalues and $C_- C_+$ will have double these eigenvalues.

To determine the eigenvalues of $A_- A_+$, let the state $|\lambda\rangle$ denote an eigenstate of this operator with eigenvalue λ:

$$A_-A_+|\lambda\rangle = \lambda|\lambda\rangle \tag{A.7}$$

We use (A.2) to derive the relation

$$[A_-A_+, A_+]_- = -A_+ \tag{A.8}$$

so that $A_+|\lambda\rangle$ is an eigenstate of A_-A_+ with eigenvalue $\lambda-1$:

$$\begin{aligned}A_-A_+(A_+|\lambda\rangle) &= A_+A_-A_+|\lambda\rangle - A_+|\lambda\rangle \\ &= (\lambda-1)A_+|\lambda\rangle\end{aligned} \tag{A.9}$$

Similarly $A_-|\lambda\rangle$ is an eigenstate of A_-A_+ with eigenvalue $(\lambda+1)$: $A_\pm$ are thus annihilation and creation operators for eigenvalues of A_-A_+. We now use that $\langle\lambda|A_-A_+|\lambda\rangle$ is always positive, since evidently A_- is the complex conjugate operator to A_+. Then

$$\langle\lambda|A_-A_+|\lambda\rangle = \lambda \geq 0$$

This would be in contradiction with our finding above that $(A_+)^n|\lambda\rangle$ is an eigenvector of A_-A_+ with eigenvalue $(\lambda-n)$, for arbitrarily large n, unless for some n, $(A_+)^n|\lambda\rangle = 0$, so such an n must exist. If this is so, we may choose a state $|0\rangle$ with $A_+|0\rangle = 0$, in particular the state $|0\rangle = (A_+)^{n-1}|\lambda\rangle$. If we now consider the state $(A_-)^r|0\rangle$, this will have eigenvalue r of A_-A_+, where r is any positive integer or zero. Since only such states are possible, as we have seen, then the eigenvalues of A_-A_+ are $0, 1, 2, \cdots$, and our result on L_z is proved.

APPENDIX 2

Physical Constants

Velocity of light = $c = 3 \times 10^8$ m s^{-1}

Electronic charge = $e = 4.8 \times 10^{-10}$ e.s.u. = 1.6×10^{-19} coulombs

1 MeV = 1.6×10^{-13} joules

'Planck's constant' = $\hbar = 6.58 \times 10^{-22}$ MeV s = 1.05×10^{-34} J s

Boltzmann constant = $k = 8.6 \times 10^{-11}$ MeV deg^{-1}

Mass of electron = $m_e = 0.51$ MeV/c^2

Mass of proton = $M = 938.3$ MeV/c^2

Fine structure constant = $\alpha = e^2/(4\pi\varepsilon_0 \hbar c) = 1/137$

Compton wavelength of electron = $\hbar/m_e c = 3.86 \times 10^{-13}$ m

INDEX

Accidental degeneracy, 126
Alpha decay, 85
Angular Momentum, 106
  and rotation group, 133
  and electron, 198
  commutation relations, 107
  classical, 26, 30
  conservation of, 132
  eigenvalues, 117
  operator form, 111
  shift operators, 114
  wave functions, 118
Antisymmetric States, 158
Approximation Methods, 145, 147, 162

Barrier, centrifugal, 119
Barrier, penetration, 85
Blackbody radiation, 37
Bohr atom, 42
  radius, 43
Born approximation, 181
Bosons, 158
Boundary conditions, 79
  periodic, 99
  scattering, 174
Bound States, 71, 189
Box, one dimensional, 85

Centre of mass, 143
Central forces, 118
Classical limit of harmonic oscillator, 94, 95
Closed shells of atomic electrons, 160
Commutator bracket, 106
Commuting set of operators, for central forces, 132
Complete orthonormal set of vectors, 62
Compton effect, 40
Conservation laws, 130
  of angular momentum, 131
  of linear momentum, 135
Co-ordinate rotations, 133

Correspondence principle, 46
Coulomb scattering, 173, 182
Cross section, 173
  differential, for scattering, 172
  total, for scattering, 173
  method of partial waves, 183

de Broglie's relation, 47
Degeneracy
  accidental, 126
  fundamental, 120
  in perturbation theory, 153
Delta function, 60, 65
Density of states, 150
  of radiation in a box, 37
Differential cross section, 172
Diffraction uncertainty, 69
Dirac notation (bra-ket), 115
Dirac theory of electron, 198
Displacement operator, 135
d-states, 121

Effective range, 193
Eigenfunction or eigenvector, 59
  for angular momentum, 118
  for harmonic oscillator, 93
  for hydrogen atom, 126
Eigenvalue, 59
  for angular momentum, 117
  for harmonic oscillator, 93
  for hydrogen atom, 125
Electron spin, 157
  relativistic discussion, 198, 199
Energy conservation, classical, 18
  quantal, 130
Energy levels, harmonic oscillator, 93
  hydrogen atom, 125
  periodic potential, 100
Equations of motion, Newtons, 16
  Hamilton's, 23
  Lagrange's, 27
  Schrödinger's, 51, 52
  Heisenberg's, 129

INDEX

Exclusion principle, 159
Expectation value, 53, 54, 63

Fermions, 158
Field Theory, 200
Forces, external, classical, 29, 30
　quantal, 141
Fourier integral theorem, 60
　transform, 60
Free particle, 51

Gamma-ray microscope, 69
Gaussian wave packet, 68
Generalised co-ordinates, classical, 26
Generalised momentum, classical, 27
Generators of rotation group, 133
　　　　of translation group, 135
Green's function (for scattering), 176
Group velocity, 50

Hamilton's equations, 23
Hamiltonian for harmonic oscillator, 90
　for hydrogen atom, 122
　for many particle system, 142
　for particle in magnetic field,
　　classical 24
　　quantal, 152
Heisenberg equations of motion, 129
　picture, 128
Hilbert space, 57
Hydrogen atom, 43, 121

Identical particles, 158
Impact parameter, 172
Infinite dimensional space, 57
Inner product, 58
Integral equation for scattering, 179
Invariance and symmetry, 134

Kronecker delta, 57

Lagrangian equations, 27
Laguerre polynomials, 126
Least action, principle of, 32
Legendre's equation, 114
Legendre functions, 121
Linear operator, 59
Linear vector space, 58
lth partial wave cross section, 185

Magnetic field, 24, 152
Many-particle system, 158
Matrix mechanics, 129
Metastable state, 194
Momentum, classical, 16, 28
　quantal, 51
Motion of wave packets, 73

Nodes, in radial wave functions, 121

Observable, 53, 63
　Heisenberg equations of motion of, 129
Orthonormal basis, 62
Oscillator, and classical motion, 21
　one dimensional simple harmonic, 90

p-states, 121
Parity, 135
　and spherical harmonics, 136
Particle flux, incoming, 174
　scattered, 175
　wavelength, 48
Pauli exclusion principle, 159
Pauli spin operators, 198
Perturbation theory, first order, 145
　time independent, 146
　time dependent, 147
　of scattering integral equation, 180
Photoelectric effect, 38
Photon, 39
Planck's radiation law, 38
Plane wave, expansion in spherical harmonics, 183
Potential in conservative field of force, 17
　step, 82
　well, 85
　general, 96
　periodic, 97
Probability current, 72
　density, 72
Propagation during scattering, 180

Quantisation, of a classical system, 51
Quantum Mechanics, basic postulates, 63

Radial wave equation, general potential, 119
　hydrogen atom, 123
Reduced mass, 144

INDEX

Reflection coefficient, 80
Relative co-ordinates, 143
Resonance, 195
Rotation operator, 133
Rutherford cross section, 173, 182

s-states, 121
Scattered wave, 176
Scattering amplitude, 174
  Born approximation, 180
  by square well, 186
  integral equation, 179
Schrödinger equation, 51, 52
  and bound states, 71
  and scattering, 176
Schwarz's inequality, 57
Selection rules, 156
Separation of variables, 118
Spherical harmonics, 118
Spin, 157, 198
Square integrable wave functions, 54
Stability of bound states, 71
Symmetry, 132

Symmetry character of wave function, 158
Systems of particles, 158

Time dependent perturbation, 150
Total scattering cross section, 173
Transition probability, 151

Uncertainty principle, 67
Unitary operator, 129

Variational principle, 162

Wave function, antisymmetric, 159
  harmonic oscillator, 93
  radial, 119
  square well, 89
  time dependent, 52
Wave packet, 49
  classical motion of, 73
  gaussian, 69

Zeeman effect, 151
  anomalous, 156